The Economic Competitiveness of Renewable Energy

Scrivener Publishing
100 Cummings Center, Suite 541J
Beverly, MA 01915-6106

Publishers at Scrivener
Martin Scrivener (martin@scrivenerpublishing.com)
Phillip Carmical (pcarmical@scrivenerpublishing.com)

The Economic Competitiveness of Renewable Energy

Pathways to 100% Global Coverage

Winfried Hoffmann

Scrivener
Publishing

WILEY

Co-published by John Wiley & Sons, Inc. Hoboken, New Jersey, and Scrivener Publishing LLC, Salem, Massachusetts.
Published simultaneously in Canada.

For general information on our other products and services or for technical support, please contact our Customer Care Department within the United States at (800) 762-2974, outside the United States at (317) 572-3993 or fax (317) 572-4002.

Wiley also publishes its books in a variety of electronic formats. Some content that appears in print may not be available in electronic formats. For more information about Wiley products, visit our web site at www.wiley.com.

For more information about Scrivener products please visit www.scrivenerpublishing.com.

Cover design by Russell Richardson

Library of Congress Cataloging-in-Publication Data:

ISBN 978-1-118-23790-8

Printed in the United States of America

10 9 8 7 6 5 4 3

Contents

To our grandson Elija
and all other grandchildren around the globe
for a future worth living.

Foreword

by Hans-Josef Fell

The time has come for an immediate change to 100% renewable energy sources on a global scale.

Even though renewable energy is currently on everyone's mind, this was not always the case. Or was it? In earlier times renewable energy was also explored. Examples of this are the 1939 Spanish solar stocks and Rudolf Diesel's first diesel engine that ran on cold-squeezed vegetable oil. Also, as early as the 19th century there were tens of thousands of wind energy systems watering agricultural fields. Even Werner von Siemens was convinced that photovoltaics would conquer the world.

However, cheap oil, together with coal and later gas and uranium, increasingly replaced the utilization of renewable energy sources with the following disastrous consequences:

- Huge areas of Japan, White Russia, and Russia are permanently destroyed by radioactivity.
- The global mean temperature has been increasing in – as seen from a history of the earth point of view – a breathtaking velocity above 0.8°C compared to the preindustrial level. The consequences are already disastrous: droughts and crop failures, flash floods and rising sea levels, hurricanes, typhoons, and tornados. All in all these have already left a deadly path on a large scale. Global warming has already destroyed lives and resulted in the flow of refugees. The global loss of species has reached an unacceptable level. A further warming up to 2°C – only an additional 1.2°C to the above mentioned value – cannot be tolerated, even if many do not believe it to be reachable.
- The global economy is increasingly being challenged. The state of indebtedness for many is on the rise, not only because

of increasing subsidies for fossil energy systems, but also due to the ever-increasing burden of rising fossil energy prices. In one year alone, from 2010 to 2011, the European oil bill increased from 280 to 400 billion US$. And from 2009 to 2010, in order to keep energy prices at a reasonable level for the end users, the global subsidies for fossil energy increased from 312 to 409 billion US$.

- Exceeding the global maximum oil extraction (peak oil) not only leads to a dramatic increase in the price of oil, but also to conflicts over ever diminishing fossil fuel resources. That is why civil wars mainly occur in oil rich countries such as Nigeria, the Sudan, and Venezuela. It is not surprising, therefore, that the oil rich region in the Middle East became a tinderbox that disrupted global peace. The two wars in Iraq, among others, were also fought for oil.

- Even as oil company profits are increasing immeasurably due to rising oil prices, poverty is increasing in the world.

Renewable energy can, and must, make a major contribution towards positive global solutions that address these crises. Renewable energies are decentralized and create jobs that provide earnings for billions of people and not for only a few companies. They relieve energy users from the ever-increasing price of energy raw materials. They make a major contribution to global peace simply because a war will not be fought over sunrays or wind. They are able to stop global warming because they are emission free, or emission neutral in the case of sustainable growth of bioenergy. Together with active carbon sinks, with safely stored atmospheric carbon as "humus" in the upper levels of soil, one can even organize a cooling effect for the earth.

This global change towards 100% renewable energies could evolve quickly if politicians and financial managers create appropriate framework conditions. If private capital was profitably oriented towards renewable energies instead of climate destroying fossil energies, there would be massive investments in them. This was demonstrated in the year 2000 with the feed-in tariff of the renewable Sources Act in Germany. In only one decade employment in the renewable energy sector in Germany increased from 30,000 in 1998 to 380,000 in 2012. This was a crucial contribution to the strong economy of Germany. In the same decade there was an increase for PV to the electricity supply from 0.01 to 8% in the industrialized country of Bavaria, with a still-growing investment velocity. What is possible in Bavaria is also feasible globally. This is even more so because solar electricity as well as electricity from wind and hydropower costs less to produce today compared with new conventional power plants.

The industrial revolution for information technology was initiated at the California universities of Stanford and UC Davis. In less than 30 years personal computers have conquered the world, even though in 1988 the head of IBM still believed that there would be no personal computers at all. At the same universities a global plan was developed which showed that a complete change of the global energy supply to 100% renewable energies by the year 2030 is industrially and technologically feasible. Furthermore, scientists Jacobsen and DeLucchi demonstrated that this would be considerably less expensive compared to persisting with the conventional energy supply.

There will be losers in this process such as the conventional economy based on oil, gas, coal, and nuclear energy. However, the global community should no longer consider their economic interests. The fight against poverty, the empowerment of economic development instead of state bankruptcies, and the securement of global peace and effective climate protection are much more important than the profit interests of the oil, nuclear, and coal industries. A global change towards 100% renewable energies is possible, giving mankind the chance to quickly change any ruinous developments for the better.

Former Member of the German Parliament,
Former spokesperson for energy and technology for the party
"Bündnis 90 / Die Grünen"
Vice president of Eurosolar
December 2013

Preface

It is the pretension of this book to give a comprehensive picture of today's energy world, to describe the potential for energy savings which can be achieved and to get an understanding of technology development which will lead to a 100% renewably powered world as the most likely situation. This is based on the long-term economic and ecological superiority of renewables over traditional energy sources. It is the combination of these topics which makes the book unique. This abstract can also be used by the reader to make his or her own sequence for the 11 chapters according to personal preference – although for those who are no experts in the field it is useful to follow the given order.

The **Introduction (Chapter 1)** starts with a description of a general phenomenon, namely the fundamental changes taking place in the world we are living in today, not only in the area of energy but also the way in which we communicate and exchange information across the globe. It critically analyzes what the general public is told by incumbent political and industrial institutions. It is this change which lays the ground for a major alteration including in how to use and produce our daily energy - from centralized systems back to municipal and even individual levels. Environmental concerns and the growing awareness about the finiteness of traditional and affordable primary energy (fossil and nuclear) will accelerate this change.

The **Analysis of Today's Energy Situation (Chapter 2)** first describes basic energy terms for those who are not experts in this field. Today's global energy situation is analyzed in form of primary (140 PWh), secondary (90 PWh) and end user energy. The various energy sectors – mobility, industry and private/office/SMEs (small and medium enterprises) – where the secondary energy is used, are specified.

When analyzing the potential of the various exhaustible energy sources and simply comparing the total sum to our current energy usage per year, it is tempting to simply divide these two numbers. The result varies – depending on how many unconventional sources are considered – between many centuries and up to two millennia. It is emphasized that a much more differentiated look must be taken.

There are major challenges for fossil energy: finiteness leading to "peak oil and gas" in the foreseeable future will result in increasing prices and CO_2-emissions, which either will add significantly to the generation cost (if carbon capture and storage can be realized) or will cause irresponsible temperature increases due to Greenhouse Gas emissions. A journey through the history of our earth dramatically outlines what we are currently doing by passing the CO_2-concentration of 400 ppm due to burning fossil fuels. Such a level was only seen millions of years ago. Accelerating factors of global warming like the melting of glaciers and permafrost thawing are also described and add to the imperative: we must ACT NOW!

The problems with nuclear technology for all reactor types, fission, fast-breeder and fusion are detailed and it is shown that these problems simply add to the inferiority of this type of electricity production; an inferiority which already stems from their higher generation cost compared to renewables – assuming no subsidies for either technology.

The **Importance of Efficiency Measures (Chapter 3)** has nothing to do with renewable technologies but the measures described will help a lot to achieve a 100% renewable contribution sooner rather than later. The importance for the general public to understand that efficiency measures will not decrease their comfort of living but will give them the "same quality of life with much less energy" is highlighted. In conjunction with renewables as secondary energy provider this will even change to a "better quality of life with much less energy".

Emphasis is placed on the fact that many energy efficient products have a higher price when purchased compared to older ones which, however, is more than offset, when the total life cycle cost or levelized cost of service is considered as shown through the example of future lighting. The superiority of electro mobility will only evolve if electricity is provided by renewable technologies. The significant savings in heating and cooling energy for houses will be seen with well insulated houses in the future, which will in parallel lead to less solar thermal appliances. As a summary, the secondary energy which would be required today with the available energy efficiency measures (~45 PWh) within the various sectors is given. A recommendation to politicians is provided on how best to accelerate the introduction of energy efficient products, namely through reasonable support for the new technologies and not simply by banning the old ones.

An **Overview of the Most Important Renewable Energy Technologies (Chapter 4)** starts with an outline of the huge potential of renewable energy resources. The outstanding offering from solar irradiation exploited by three technologies – decentralized PV, centralized concentrated PV and solar thermal electricity (or concentrated solar power, CSP) as well as solar thermal low temperature – is highlighted. Simply considering land use and near-shore coastal regions for wind off-shore ("technical potential"), we can provide 880 times today's secondary energy through renewable energy and when taking today's technologies and feasible areas ("sustainable potential"), ~35 times today's secondary energy could be provided (and 21 times the future secondary energy needs for our 100% renewably powered world).

The historical developments of technology and market are detailed for wind energy as well as for solar thermal collectors and concentrators. Readers will understand why wind turbines are getting higher and higher to make best use of the wind conditions at a given site. An overview of PV and other renewables (hydro, geothermal, wave and tidal) is also given.

The **PV Market Development (Chapter 5)** starts with a topic which is addressed to economists, in particular to liberal ones. When it comes to the question of whether support schemes are useful or whether only free market mechanisms should decide on certain technology developments it is advisable to differentiate between strategic goods – such as electricity production, transportation – and consumer goods – like mobile phones or televisions. It is shown that while for the second group the free market mechanism is the right instrument, this is fundamentally different for the strategic goods. When it comes to the question of whether support for a technology should be organized through market pull or technology push, the clear answer from an industrial point of view is through market pull.

The development of a multitude of different customer needs for PV products and the associated market volume is discussed from the 1970s until today. The unimaginable average market growth of more than 50% per year in the first decade of this century was only possible due to the support scheme in form of the Feed-in tariff, where renewable technologies went along with a long-term payment (typically 20 years) for all produced electricity based on the respective cost plus a positive margin. Based on the ideas by Wolf von Fabeck and municipal experience in Switzerland, it was Hans-Josef Fell and a good number of supporters who first got it politically up and running in Germany, after which it spread out into more than 60 countries worldwide. The total budget for these payments (minus the stock exchange value) can be seen as an investment by society and it is shown that the associated Net Present Value is clearly positive with conservative assumptions. Even if there is considerable outcry over the many billions spent on this investment, it obviously pays off when analyzed over the long-term payment period. The fact is that after this time period our children will benefit from a "golden age" in which electricity is produced at marginal cost with depreciated PV systems throughout their life-time which is significantly longer compared to the typically 20 year's payment time period.

In times of high annual growth as mentioned above, some bottle-necks along the value chain (e.g. poly silicon) appeared, resulting in an increase in prices. This was taken as a signal to invest in additional production capacities all along the value chain. With clear industry political goals coupled with a number of clever entrepreneurs it was Asia, particularly China which increased its global share in PV module production from about 5% in 2005 to 60% in 2012. Unfortunately the capacity increase outgrew the market volume which resulted in about 100% overcapacity in 2012. As in every industry the consequence is now a shake-out of production companies associated with (too) low product prices and deep red numbers on the balance sheet. The flipside of this situation is that it allows new markets to establish themselves which would not have been realistically possible only a few years ago. After this consolidation period and a further market growth we will see a new wave of production facilities which, with new ideas from the R&D-institutes, will enable cost numbers which are low enough to achieve positive margins at today's prices. In 2013 we are in a time where we clearly foresee the end of the running Feed-in tariff program in only a few years, which leads to the necessity to install a new market design for the future increased levels of renewable electricity including the procedure how renewable electricity is traded on the stock exchange. The development of electricity storage, Demand Side Management, smart grids and virtual power stations is described.

The **PV Value Chain and Technology (Chapter 6)** summarizes the various PV technologies c-Si wafer and Thin-Film in greater detail, but also describes concentrated PV, Dye solar cells and organic devices. Besides modules, the additional components for a complete PV system such as inverters and BOS (Balance Of Systems) components are also dealt with. Based on a number of examples an important observation is described: the power of continuous development and economy of scale versus breakthrough technologies to decrease production cost is most often underestimated.

The Astonishing Predictive Power of Price Experience Curves (Chapter 7) shows impressively what even research and industry people from the same technology sector could often not believe. Such curves plot the cumulated volume of a particular product versus the respective price in a double logarithmic scale. From the slope one determines the %-change in price for each doubling of cumulative volume. With the example of DRAM semiconductor devices it is demonstrated that all people strongly believed in the 1970s/1980s that the slope was horizontal – i.e. no further price decrease – after the 1990s. Yet 20 years later we are still running down this same graph. Similarly for Flat Panel Displays such a development has been on-going since the 1990s.

Emphasis is placed on the fact that while big and centralized technologies have more of a project character for specific regions, small and mass produced components which are globally produced and internationally traded have a high probability of reaching significant cost and price reductions. Examples of centralized technologies are power stations; examples of mass produced components are PV solar modules, batteries and fuel cells.

The Price Experience Curve for solar modules (c-Si and Thin-Film) is shown as well as for inverters and by analogy to the examples given before, there is sound reason to believe in a further continuation of falling prices. Deviations from the slope can be explained by e.g. product shortage for prices above and overcapacity for prices below the Price Experience Curve.

The **Future Technology Development (Chapter 8)** is discussed for all renewable technologies. The potential price development is based on the respective Price Experience Curve

For **Future Energy Projections – The 150 PW-Hour Challenge (Chapter 9)**, some well-known projections from the IEA and Greenpeace are shown for reference. The market development for the various renewables and their potential market share are described in some detail. Reasoning for a simple split to cover the required 150 PWh of energy for future secondary energy is explained: decentralized PV, centralized concentrated PV and solar thermal electricity systems, decentralized low temperature heat, wind energy and the rest of all other renewables provide 20% each, or 30 PWh of energy.

Realizing that if renewables will take over the 100% energy supply there will be huge industries for all technologies and therefore a great opportunity for all economic regions to grow a sizable industry for this future. Individual companies should be encouraged to make an extra effort to be in that business. With the example of PV it is estimated that the annual overall turnover in the 2040s will be comparable to the global annual turnover for the automobile industry.

The **Likelihood and Timeline for a World Powered by 100% Renewables (Chapter 10)** deals with the potential development. While some years ago the need for a global network in form of a worldwide super grid was seen as an elegant solution, a new model may emerge: local autonomy for the decentralized private and SME sector and relocation of the energy intensive industries, which need power and process heat, close to places which cost effectively deliver energy from the centralized hydro, (concentrated) PV and (off-shore) wind parks (sometimes it may be more cost effective to link power intensive industries via transmission lines to the big renewable power stations).

In **Conclusion: The 100% Renewable Energy Puzzle (Chapter 11)** summarizes the findings and discussions from this book in form of a 3 by 3 element puzzle. The limit of a 2°C temperature increase can only be accomplished if we shift all fossil power stations to renewables as quickly as possible. Nuclear is not a viable alternative for safety and cost reasons. Driven by mass produced products like solar modules, batteries and other devices which are important for the future 100% renewable world the cost and price decrease will demonstrate the economic superiority of renewables over traditional fossil and nuclear technologies by the 2020s at the latest. Once this is recognized by the financial community there will be a substantial re-allocation of huge investment money, away from traditional technologies and a steady move towards the 100% renewably powered world.

Acknowledgements

This book would never had been possible without the continued support and the silent agreement from my wife Anneliese not to pursue her own career but to care for the family and guide two great kids, Tobias and Elisabeth, from birth through childhood and school into their adult age.

Only this allowed me to devote my time – not only during business hours – to growing a company (ASE – Applied Solar Energy) from scratch to one of the five largest cell and module producers in the early years of the new millennium. My thanks go out to all who supported me in those years, from colleagues and employees to shareholder individuals like Heinz-Werner Binzel, who believed in this new business and supported it in an environment where one would not always have expected it. My thanks go out to all customers who dared to buy the new products in those early days. Equally vital was the cooperation with machine builders and scientists, at which point I wish to highlight Prof. Rudolf Hezel from ISFH and Prof. Joachim Luther from the Fraunhofer-ISE.

At the same time, it was necessary to grow a market with the engagement of colleagues from other companies, who in the end turned out to become good friends, like Hubert Aulich, Tapio Alvesalo, Günther Cramer, Georg Salvamoser and many others in associations like BSW Solar and EPIA, who convinced and supported politicians in creating market support programs – the latter group is best represented by Hans-Josef Fell, to whom I am also grateful for the Foreword of this book. I also enjoyed the cooperation with other renewable associations which gave me the chance to get a better understanding of the other renewable technologies – this was a prerequisite for delving deeper into these products and into the important role which they play.

Not being a native speaker, I very much enjoyed how my German English was shaped into English English by Jenny Taylor. Last but not least it was Martin Scrivener who convinced me to put this information into a book after a plenary talk I gave in Chicago at the start of a big international scientific conference back in spring 2011(!) on this very subject. Had I known from the start how much pain it would take to bring this to a successful conclusion, I do not know whether I would have even started. But in the end I found it to be worthwhile and I thank Martin for his continued encouragement which helped me, together with the constructive comments from my students on the topics, to complete this book.

List of Abbreviations

Abbreviation	Full name (translation or additional information)
BAPV	Building Added Photovoltaics (modules attached to existing buildings)
BIPV	Building Integrated Photovoltaics (modules as integral part of buildings)
BSW-Solar	Bundesverband Solarwirtschaft (German Solar Economy Association)
BMU	Bundesministerium für Umwelt, Naturschutz, Bau und Reaktorsicherheit (German Federal Ministry for the Environment, Nature Conservation, Building and Nuclear Safety)
CEO	Chief Executive Officer (Head of management board of a company)
CCGT	Combined Cycle Gas Turbine
CCS	Carbon Capture and Storage (also called CSS = carbon sequestration and storage)
CHP	Combined Heat & Power
DEEM	Direct Energy Equivalent Method
DSO	Distribution System Operator (Medium and low Voltage)
EEG	Energie Einspeise Gesetz (Renewable Energy Sources Act, FiT in Germany)
EPIA	European Photovoltaic Industry Association

EREC	European Renewable Energy Council (umbrella organization of all major European renewable technology associations)
FiT	Feed-in Tariff
GHG	Green House Gases (CO_2, CH_4, water vapor etc.)
IEA	International Energy Agency
IPCC	International Panel on Climate Change
OECD	Organization for Economic Co-operation and Development
PE	Primary Energy (Energy embodied in natural resources)
PV	PhotoVoltaic
P2G	Power to Gas ((renewable) electricity → Hydrogen by electrolysis + CO_2 → CH_4)
R&D	Research and Development
RE(S)	Renewable Energy (systems)
SE	Secondary Energy (transformed Primary Energy, also called final Energy, which reaches the final consumer's door plus energy used in the energy sector itself, e.g. transmission losses for electricity)
SKE	Stein Kohle Einheit (hard coal equivalent)
SME	Small and Medium Enterprises
TSO	Transmission System Operator (High Voltage)
UNEP	United Nations Environment Program
WBGU	Wissenschaftlicher Beirat der deutschen Bundesregierung für globale Umweltveränderungen (Advisory Board to the German Government on Global Change to the Environment)
WEO	World Energy Outlook (by IEA)
WMO	World Meteorological Organization
WTO	World Trade Organization

1

Introduction

1.1 The Changing World

We are living in a world of fundamental changes: since industrialization started with the steam engine in the 1770s by James Watt we have experienced phenomenal growth in many areas. Let's take the global population as an example, which grew from 1 billion in 1800 to 3 billion in 1960 and to more than 7 billion today. Energy consumption grew from 3 PWh (10^{12} kWh) in the mid 1800 to ~150 PWh today, CO_2 emissions grew from ~0.2 Gt (billion tons) in early 1800 to ~30 Gt (fossil and other) today. The mentioned increases between 1800 and today are an impressive factor of ~7, ~50 and ~150 for the global population, energy use and CO_2 emissions, respectively.

Until the 1970s, there was only little concern over whether there should be any worries about the finiteness of resources. This changed after the first serious publication on energy and material scarcity in form of the report "Limits to growth" by the Club of Rome in 1972 [1-1] and, coincidentally, the first oil price shock in 1973. Until recently, only concerned individuals and dedicated organizations highlighted the finiteness of traditional primary energy sources – now the IEA (International Energy Agency) points to the same fact in their latest World Energy Outlook. An atmosphere of change has evolved, due to concerns over climate change which will ultimately have a dramatic impact on the human population and which is caused by increased CO_2 concentration mainly by burning fossil sources to produce electricity.

What are often called the "residual" risks associated with nuclear power should be renamed (as in Germany) "intolerable" risks as has been demonstrated again in Fukushima but also due to not manageable potential terrorist actions. The unsolved storage issue of spent fuel for nuclear reactors is of additional concern. After all, it is also the increasing cost figures which do not even contain insurance numbers as no insurance company worldwide is willing to take on the associated risk.

Of course, current industries which have established themselves well in the old environment do not like such developments and are trying to find ways to at least prolong the business as usual scenario. Together with their closely attached supporters from politics and industry they are instead trying to push for solutions which they themselves could most easily realize: CCS (Carbon Capture and Storage) for fossil use and new nuclear reactor types like Fast Breeder and Fusion. There is, however, one very clear fact: you cannot stop a better product whose time has come to replace the old one. And those who are not willing to adapt will not survive – as will also happen in the energy sector. In contrast, a multitude of new players entering the energy field very often underestimate the challenges – at least time-wise – which are linked to the introduction of more and more decentralized renewables, new customer behavior, energy storage, energy efficiency measures and much more.

The large scale introduction of renewable energies also needs better communication at the local level since they demand much stronger decentralized solutions. This process of decentralization has already taken place in many industries. Let us take a look to some examples from the past with first, the introduction of personal computers in the 1970/1980s. Initially, most experts focused on the large centralized supercomputers, whereas by now most of the computing power has been taken over by hundreds of millions of decentralized personal and company based computers, while in a few areas such as global climate models, large supercomputers are still needed and running. Another example is the widespread use of mobile phones which happened just in the last 20 years. Again, had someone told the experts from the communication industry in the late 1980s that in 2010, an individual would be able to reach almost anyone anywhere on the globe with a device as small and light as a cigarette box, they would have been declared stupid and lacking in technological understanding. However, as we all know, technological development told a different story. This is very similar to the question of whether we will have millions of decentralized energy producing systems, based on renewables which do not need any exhaustible fuel and do not harm the environment or whether, as the energy experts of today are telling us, only big power stations will provide the energy we need at a lower cost – but in most cases only if the external costs are not internalized. This book will provide arguments for the much higher probability of the renewable solution.

1.2   Why Another Book on 100% Renewables?

A large number of books and studies have been published in recent years on how to change the old world into something new which would overcome the above mentioned challenges. It is well known that projections into the future are very difficult to make and one should not dare assume that a specific and detailed model will really develop exactly as predicted. This is true for all future projections, be they on future climate development, energy portfolios, population growth or other. Unfortunately, in many cases all the different scenarios have to be evaluated at the same time, since the result of one, e.g. assumed population at a later time, dictates how much total energy will be needed by then, which in turn will have an impact on future climate development predictions. Another difficulty arises when in some studies the reader can no longer see the "red thread" in the story because it is buried and hidden among too many meaningless 3-digit numbers.

Everyone is biased – and so am I. Nevertheless I try in the best possible way, being a scientist by training, to understand various alternatives and to base my future expectations on a solid analysis of the past, if applicable. Having been interested in the development in ALL renewables for a long time and in particular in technology and market development for PV, I take great pleasure in passing this knowledge on to young students in university lectures since my retirement from the operational business. This is the basis for this book which does not aim to elaborate the details, but to find out which developments may come about with higher probability compared to others. Having had the privilege of working in a number of different fields I was able to build a fairly solid knowledge base.

This book is about the shift in paradigm from traditional energy to 100% renewables. Throughout my work life, I have experienced how difficult it is for the traditional energy people to follow the extremely quick developments in the renewable field which is not a surprise. Many decades of experience have shown that changes in the traditional fields only happened very slowly and in most cases in the long run these changes always had a higher cost and price as a result of increasing fuel prices (and various other parameters). Traditional actors simply cannot understand how technology development for wind and PV can lead to such a quick decrease in cost and price as we have seen just over the last decade. Unfortunately this misunderstanding is also shared by most politicians, which leads to problematic decisions on the legal framework. During the preparation of this book over the last two years, I personally experienced how technology developments can – or rather: should – change one's mindset. As an example, only two years ago I was strongly convinced that we need a global super-grid to balance the variable renewables with the local load and energy needs in order for our energy to be provided 100% by renewables. A better understanding of how the load duration curve can be balanced, and the quick development of storage solutions from small (kWh-range, also driven by automotive development) to very large ("Power to Gas") has cleared the way for more and more opportunities to produce all energy required in a particular region within the same local area, which can be defined as many Low Voltage Smart Grid regions with an overlap to adjacent ones. Energy intensive industries will locate themselves in areas with high levels of renewable power generation like hydro, off-shore wind and very large PV and solar thermal electricity producing plants. Alternatively transmission lines can transport electricity form the large renewable power stations to the respective industries.

Readers who are interested in how the future world can and most probably will be energized will get an easily understandable summary. Simple facts based on "rule of thumb" and "order-of-magnitude" considerations provide a basic understanding for the "total and global" picture. A wealth of detailed reports is available from many organizations to describe future projections on energy needs and how to meet these needs. Quite often, these studies follow the interest of traditional pathways and narrow-mindedly neglect and even deny the potential contribution of renewables. In more than 30 years of working in the renewable Energy sector, I have successfully helped to shape the development of the photovoltaic industry. Firstly as CEO of ASE (later renamed RWE Solar), which in the late 1990s was one of the five leading production companies for solar cells and modules (SCHOTT acquired this company in two steps, 50% in 2002, 100% in 2005 and, unfortunately, just closed the PV business at the end of 2012). Secondly as board member of the two PV associations BSW Solar (German Solar Economy Association) and EPIA (European PV Industry

Association), continuously fighting and arguing for the prosperity of the PV industry and market.

This book does not pretend to provide detailed quantitative numbers on future energy scenarios. It is rather an attempt to describe qualitatively a – maybe the most – realistic development in energy usage and production, based on well-known technology based developments, such as semiconductors, flat panel displays and more.

I am fully aware that future predictions, especially when looking many decades ahead, are always associated with increasing error bars. In addition, the anticipated starting framework conditions in many cases determine the end result. Based on my industrial experience, this book gives a set of assumptions and by simply extrapolating known technology developments it will be shown that a quick 100% coverage of the global energy needs is much more probable than the most often discussed case of "business as usual".

The world is at a crossroads between either changing the energy picture towards the efficient implementation and use of renewable energy sources very quickly, or postponing this process by approximately 30 to 50 years which corresponds to the lifetime of one additional investment cycle in traditional energy systems. This latter case would significantly add to the energy and environmental cost compared to the quick renewable route. Simply for monetary reasons, a third alternative, namely the use of traditional energy sources like fossil and nuclear in the long term future is highly unlikely.

2

Analysis of Today's Energy Situation

2.1   Basic Energy Terms

Energy, measured in Joule [J], is the product of power, measured in Watts [W], multiplied by time, measured in seconds [s]. In technical terms, it is convenient to measure time in this context not in seconds, but in hours (Wh, which is 3,600 Ws = 3,600 J). Through using a prefix as shown in Table 2.1, the huge span of different energy contents can be described. The table also shows the prefixes for small dimensions, often used for length (measured in meters, m), weight (measured in grams, g) and time (measured in seconds, s, also needed in later sections). Except for the very first and the last interval there is always a factor of 1,000 separating the various prefixes. Starting from the unit, each prefix for smaller numbers is one thousand's of the preceding: 1 milli = 1/1,000 unit, 1 micro= 1/1,000 milli and so on. Similarly, for increasing numbers, each following prefix is 1,000 times larger than the preceding one: 1 kilo = 1,000 units, 1 Mega = 1,000 kilo and so on. In order to have a better feeling for the huge span ranging from 10^{-35} up to 10^{30} which covers 65 orders of magnitudes some examples

Table 2.1 Prefixes for (very) large and (very) small numbers, the scientific notion (in brackets the logarithm) together with some examples.

	prefix	Scientific Notion (log)	Examples
		10^{-35} (-35)	1.62×10^{-35} m = Plank's length
z	zepto	10^{-21} (-21)	zm (size of hypothetical preon particles as subcomponents to quarks and leptons)
a	atto	10^{-18} (-18)	am (~size of quarks and electrons)
f	femto	10^{-15} (-15)	fm (~scale of atomic nucleus)
p	piko	10^{-12} (-12)	25 pm = radius of hydrogen atom
n	nano	10^{-9} (-9)	ns, nm
μ	mikro	10^{-6} (-6)	μs, μg, μm (~50μm = human hair thickness)
m	milli	10^{-3} (-3)	ms, mm, mg, mW
		1 (unit) (0)	1s(econd), 1m(eter), 1g(ram), power:1W(att), energy: 1J(oule) = 1Ws
k	kilo	10^{3} (3)	1kg, 1km, 1kW, 1kWh
M	Mega	10^{6} (6)	1Mm = 1,000 km
G	Giga	10^{9} (9)	1.39 Gm = radius of the sun
T	Tera	10^{12} (12)	5.9 Tm = distance Pluto to sun
P	Peta	10^{15} (15)	9.46 Pm = one light year
E	Exa	10^{18} (18)	1.9 Em = 200 light years
Z	Zetta	10^{21} (21)	1 Zm (~ diameter of milky way galaxy = ~100,000 light years)
		10^{30} (30)	5×10^{30} m = diameter of universe

are given in terms of length. Although this exercise is trivial for scientists it may be helpful for others.

While the smallest dimensions equal the subcomponents of the constituents of atoms, the bigger dimensions describe the size of our Milky Way galaxy. The biggest dimension in reality, the size of our universe, is about 5×10^{30} m (or 5 billion times the size of our Milky Way galaxy) and the smallest dimension is about 1.6×10^{-35} m, the so-called Plank's length. Smaller dimensions do not make any physical sense as theoretical physicists tell us (according to their string theory). Right in the middle (micro to Mega) of the two extremes of this length scale is what we as human beings normally experience. This coincidence of why the world of human experience is around the logarithmic middle of the minimum and maximum of reality may be a topic for philosophers to speculate about.

In order to get an understanding of what the meaning of some energy contents are, Table 2.2 gives some examples. Throughout this book energy will always be described in Wh. For those who want to compare these numbers with other measurements from different publications, a conversion table is given in Table 2.3.

There exists a variety of different energy forms: primary, secondary and end user energy. The first is the energy content of the primary resources like coal, oil, gas or uranium just after mining or drilling. For convenient usage, these primary resources have to be converted to secondary energy forms. Crude oil for example is converted into the secondary energy forms diesel and petrol with fairly small losses compared to those associated with converting primary resources (coal, oil, gas or uranium) into the secondary energy electricity.

A note for the specialists: Today there are three different ways to measure the primary energy (PE) for the various constituents, (1) the Physical Energy Content Method, (2) the Direct Energy Equivalency Method and (3) the Substitution Method. The *first method*, used among others by the OECD, IEA and Eurostat measures the useful heat content for all fossil and nuclear materials as well as geothermal and solar thermal electricity

Table 2.2 Commonly used units and some examples in the electricity and energy business.

Example	power	x time	= energy
Light bulb	100 W	10 h	1 kWh
Otto engine (1 car)	100 kW	10 h	1 MWh
Electricity for one household	~ 450 W	1 year = 8,760 h	~ 4,000 kWh
~ 250 households (small village)			~ 1 GWh
~ 250,000 households (town with ~1 million people)			~ 1 TWh
Germany's electricity			~ 600 TWh
Europe's electricity			~ 3,000 TWh
World's electricity			~20 PWh
Primary energy worldwide			~ 140 PWh

Table 2.3 Conversion table of commonly used energy measures (SKE (SteinKohleEinheit) = coal equivalent).

To From	kJ	kcal	kWh	kg SKE	kg oil equivalent
1 kJ	1	0.239	0.000 278	0.000 034	0.000 024
1 kcal	4.187	1	0.001 163	0.000 143	0.000 1
1 kWh	3,600	860	1	0.123	0.086
1 kg SKE	29,308	7,000	8.14	1	0.70
1 kg oil equivalent	41,868	10,000	11.636	1.428	1

production as PE, while for those renewables producing electricity directly, this secondary energy (SE) is defined as PE. This is obviously arbitrary as I would generally define the PE as the useful energy content for all materials and processes entering a machine or process to produce the needed SE from this PE. In this case to define the PE for renewables, we would have to take the SE divided by the efficiency of the respective process. For example if we have a 20% efficient solar module producing 10 kWh electricity the PE would then be 50 kWh which is contained in the solar power used. But convention has now established the procedure I first described which will also be used in this book – except when otherwise stated. The *second method* used by the UN and IPCC is similar to the first one with the exception that non-combustion methods like geothermal, solar thermal electricity technologies but also nuclear equate the SE electricity with PE. The *third method*, mostly used by the US Energy Information Administration (EIA) and BP, is based on the convention to equate all forms of SE (electricity and heat) with the volume of fossil fuels, which would be required to generate the same amount of SE. In this case, one not only has to postulate the assumed efficiency for this procedure, but it is obviously a tribute to the "old days", when fossil sources were seen as the most important ones.

The last step is the conversion of secondary energy into end energy, which is needed to power the actual service wanted. This is again accompanied by energy losses. Examples would be diesel and petrol to drive a car from A to B, where only about 30 % of the secondary energy is transformed into motion whereas 70% is lost as heat; electricity to power a light bulb where the old incandescent lamp converted only less than 10% of secondary energy into light intensity (measured in lumen) while more than 90% were lost as heat. Another important example is the way of comfortable housing. Firstly, we should remember what our ancestors 2000 years ago already knew, namely to position the house so as to optimally collect – or keep away – solar radiation during a year. Secondly, we should use today's technologies and products for insulating walls and windows using specially coated glass panes (low e glass) in order to minimize the residual energy needed to heat and cool our houses. How the then necessary residual energy can be provided in the most cost competitive way – solar thermal collectors with seasonal storage or heating with PV solar electricity – will become an interesting question.

2.2   Global Energy Situation

The situation of the global energy need today is shown in Figure 2.1. The left column describes the primary energy consumed around 2010 as being approximately 140 PWh which is based on IEA data [2-1,2] and using the "Physical Energy Content Method". The contribution from renewable energies, including hydropower and biomass, used to be less than about 13% for a long time, fossil (coal, oil and gas) provided the lion's share of 80% and nuclear accounted for about 7%. The split for the various energy carriers is shown in Figure 2.2. It will be the goal of this book to provide all the necessary information in order to understand that all future energy needs can be cost effectively covered through using only renewables just a few decades from now.

The second column summarizes the secondary energy sources consisting of treated fossil sources (like petrol, diesel, gas) and the convenient

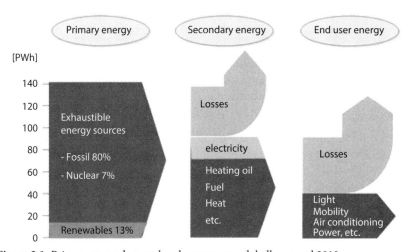

Figure 2.1 Primary, secondary and end user energy globally around 2010.

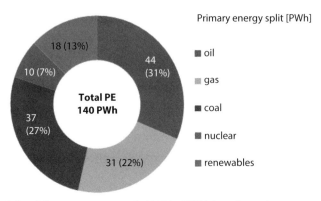

Figure 2.2 Split of the primary energy (~2010 in PWh) into the various energy carriers.

energy form electricity. The losses shown are mainly associated with converting primary energy to electricity in power stations. Although it is today possible to convert fossil fuel to electricity in a modern power station with up to ~60% efficiency, it is a matter of fact that globally the average conversion is only about one third, with two thirds of primary energy content lost as heat in the cooling towers. In the same year, the approximately 20 PWh of global electricity needs and the associated losses of ~30 PWh were obviously eating more than one third of the total primary energy (here I assumed 15 PWh from fossil and nuclear which contribute to the named losses and the other 5 PWh from hydro and other renewables with no losses). The conversion of the remaining 90 PWh of primary energy sources into 70 PWh for fuel, gas, heating oil and other secondary energy sources is associated with losses which I estimated to ~20 PWh.

The split of the primary energy between the various energy carriers is shown in Figure 2.2. In addition the fossil sector is divided into oil, gas and coal.

The secondary energy is ultimately converted into what we really need, the end energy. This is again associated with quite high losses including those for transportation and also depending on the type of end energy we are looking at. Some examples are the conversion of electricity into light, moving a car from A to B using petrol in an Otto engine, heating homes with gas or oil including heating water, and powering the various processes in industry through electricity or process heat. Only approximately 40 PWh, which is less than half of the secondary energy and less than one third of the original primary energy, is contained in the end energy. An old rule of thumb to determine how much energy would remain from process to process was the easily retained ratio of 3:2:1 for primary, secondary and end energy, respectively (today 140:90:40 PWh = 3.5 : 2.3 : 1). As we will see later, this will no longer be the case in the future. Instead, a more favorable ratio of ~1.5 : 1 for secondary and end energy can be considered due to more efficient machines and appliances.

An analysis of the timely development of the major contributors of the various secondary energy forms is very interesting, as the one conducted by Marchetti [2-3] in 1979. He found that since the 18th century when wood was the main secondary energy source, a new form gradually developed every 50 to 100 years to increasingly replacing the preceding one. Industrialization with the development of steam engines pushed the use of coal as a source of secondary energy. The discovery of the Otto engine and the quick increase in the number of cars at the beginning of the 20th century saw the development of oil as a new secondary energy source, followed by gas. Coal saw its peak at around 1920, and Marchetti's study foresaw the peak of oil to be around 1990 with gas following around 2040. An interesting finding was the strongly decreasing proportion of carbon to hydrogen content in the chemical compounds of the fuel with each transition from wood to coal to oil to gas. It is approximately 10 to 1 to 0.5 to 0.25. Nuclear, with a carbon content of zero, was still in its early stages in the 70ies and Marchetti assumed it to peak possibly at the end of the 21st century. He also made a forecast for a new secondary energy form, starting in 2020 and which he called "Solfus" an acronym for sol-ar and fus-ion. In a later publication [2-4] Marchetti elaborated more specifically on his predicted future development: gas should reach its market peak in 2040, nuclear in 2090 with a 60% market share and Solfus was reduced to fusion with a mere starting point of 1% market penetration in 2025. Only in 2090, he predicted "solar power beamed from Venus" together with elementary particle technologies. In my point of view, however, nuclear may already see its end well before the close of the 21st century and solar along with all the other renewables may take up 100% of the market share by then – what a possible and fundamental change!

2.3   Energy Sectors

The use of the three secondary energy forms is shown in Table 2.4. We will concentrate on three major sectors

- industry,
- small and medium enterprises (SME's), private homes, offices, hotels etc.
- mobility

Table 2.4 Secondary energies in main segments for 2010 (ref.: IEA [2-2], Greenpeace [2-5] and own considerations).

All numbers in PWh	electricity	heat/cool	(bio)fuel	total
Private homes, SME's, trade, offices, hotels.	11.4	25.6	–	37
industry	8.4	14.4	6.2	29
mobility	0.2	–	23.8	24
total	20	40	30	90

Table 2.5 Mobility sectors with consumed energy (PWh) and relative share (%).

Mobility sector	Annual energy consumed in PWh	Relative contribution in %
Passenger vehicles	12.8	53
Trucks	7.5	31
Ships	1.7	7
Aviation	1.4	6
Railway	0.7	3
Total	24	100

For many it is astonishing that about one quarter of secondary energy is used for *mobility* for the worldwide transportation, including private cars, buses, trucks, railway, ships and planes. Although today most mobility sectors are powered by oil we have to differentiate because the future will use different secondary energies – much more electricity, hydrogen or hydrogen based fuel (like CH_4) but little biofuel - for the various transportation vehicles.

Based on recent studies by Greenpeace [2-5] and "Quantify" [2-6] the split of the various mobility sectors within the first decade of this century was approximately as summarized in Table 2.5.

The fact that fuel for the passenger transport exceeds 50% of all fuel consumption is striking – and at the same time presents a great opportunity to shift towards renewable energy as will be discussed later. Railway transport is today the only sector which uses a significant yet still small portion of electricity (0.2 PWh) besides oil based fuel. Railway could potentially absorb a significant portion of passenger and freight transport. The areas of aviation and shipping although not highly significant in the overall fuel consumption do, however, contribute to pollution much more than the other areas. This is due to the fact that most of the planes fly at high altitudes and thereby create much secondary damage and many ships use heavy marine diesel fuel with a high content of sulfur in particular.

The sector *private homes, SME's, trade, offices, hotels and others* has a share of ~70% heating/cooling energy and ~30% electricity. It is quite surprising what a considerable portion of today's secondary energy is being used up to heat and cool houses, both private and business related ones. Mostly oil and gas is used for this purpose and only a small proportion is served through district heating using waste heat from conventional electricity producing power stations. Proper insulation could almost eliminate this huge energy consumption and is called a passive energy measure because it is not necessary to actively provide heating or cooling. It may be easy to decrease this energy need in new houses which are being built with appropriate insulation, however, there still remain the millions of old houses in inventory which will require a major effort in energy efficient renovations in the coming years. As energy efficiency for buildings is a high value topic of society, we must make the necessary investments by politically supporting measures such as low interest rates and tax deduction mechanisms. In the meantime, until all houses – including the existing ones – have this passive energy measure, it is desirable to serve as much heating and cooling energy needs through solar thermal technologies. However, in the long run this will become unnecessary as will be discussed later.

Industry satisfies most of its energy needs through process heat (~50%), followed by electricity (~30%) and fuel (~20%).

In order to easily remember the total secondary energy for the various energy sectors plus electricity, see the graphical representation in Figure 2.3. Low temperature heat and mobility take up slightly more than a quarter each, while electricity and industry (without electricity) each absorb slightly less than a quarter of the total secondary energy. Of the total electricity, about 40% are for industry usage.

One of the important drivers of inflation is the increasing price of energy, both in transportation as well as in heating (in more northern countries like Germany) or cooling (in more southern areas). For example in Germany the average shopping basket attributes 13% of expenditure to transport fuel and 31% to household gas and fuel – hence almost half of the inflation rate is bound to this increase. This could be completely changed. The depleting oil resources with increasing prices for exploration and transportation will be replaced by renewable sources. Zero emission houses of the future would no longer need fuel, which is becoming more and more expensive. Even if we do not have "zero emission" houses, "low emission" ones plus excess energy from wind and solar which could be

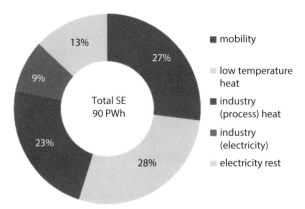

Figure 2.3 Global energy sectors for secondary energy (2010).

used to power heat pumps for heating and cooling could be even more cost effective. The annual cost for comfortable housing would remain constant, thereby suppressing inflation considerably. As food for thought, one could postulate that the replacement of exhaustible energy sources by renewable energies would change the delicate and fragile balance of today's economies towards a much more stable situation. The reason for this is that exhaustible sources are becoming more and more expensive and further technology development in renewables will even decrease their cost. I wonder how an economist would further elaborate this postulate.

Today there are many discussions on whether society is able to survive with zero economic growth. This is becoming more and more important as – hopefully sooner rather than later – the growth of the global population will also come to zero and everyone will have the same quality of life based on a fair and well defined average. Together with the fading of inflation the possibility for zero economic growth could become an interesting option for future economic models.

2.4 Challenges for Fossil Fuels

2.4.1 Finiteness of Fossil Fuels Leading to the Peak of Oil and Gas

Today's fossil fuels are the result of the accumulated growth of trees and plants over millions of years in pre-historic times which were subsequently transformed into coal, oil and gas. In its Flagship-Report [2-7] WBGU has collected all available information about the existing content for these fossil resources. For each of the resources an effort was made to differentiate between conventional and unconventional technologies to extract the individual resource from its natural surroundings and also between reserves, resources and further deposits where the individual resource is located. In Table 2.6 these data are summarized; for convenience the respective data for the nuclear resources are also added.

Bearing in mind that the annual primary energy for today is 140 PWh one is tempted after a quick look to Table 2.6 to assume that we must not worry about our future energy security: with the overall total of more than 1 million PWh we could supply our annual primary energy needs for the coming 7,000 years. However this conclusion would be totally wrong. If we plot the numbers from Table 2.6 in a graph as done in Figure 2.4 it quickly becomes evident that the biggest contributions are very questionable. Many occurrences, especially for the unconventional methods and further deposits, are either not available or only with substantial risk (examples are fossil sources below the ice shields of the Arctic and Antarctic, the extraction of methane from deep-sea methane hydrate or the use of fast breeder reactors).

I am always astonished when reading about new findings of oil and gas fields in newspapers (like the unconventional shale gas discoveries in recent years) and by the impression that is given that this would be the solution for the future. There is one very clear fact: no matter how many additional oil and gas fields we may find, there is a limited number of fossil fuels available for mankind which should not be burned but should be

Table 2.6 Global occurrences of fossil and nuclear sources [2-7].

All data in [PWh]		Reserves	Resources	Further deposits	total
Conventional	Oil	1,778	1,391	–	3,169
	Gas	1,680	2,251	–	3,931
	Coal	5,880	132,200	–	138,080
	Nuclear	672	2,072	–	2,744
Unconventional	Oil	1,064	9,520	13,160	23,744
	Gas	11,900	15,820	137,200	164,920
	Coal	–	–	–	–
	Nuclear	–	1,148	728,000	729,148
total		22,974	164,402	878,360	1,065,736

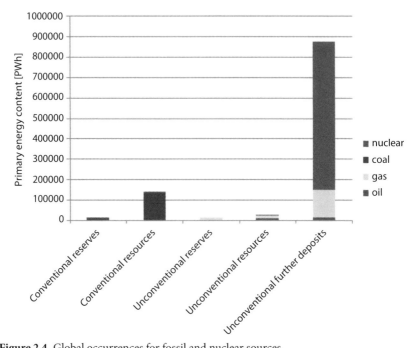

Figure 2.4 Global occurrences for fossil and nuclear sources.

kept for organic chemistry and related products for future generations. It is a shame that the arguments from many high ranking politicians and industry leaders are only guided by short term profit thinking and not by the welfare of future generations. I sometimes wonder whether these people have no kids or grandchildren.

Let us now take a closer look at the situation of oil. There are the conventional oil fields which have been established over the last decades. From Table 2.6 the known reserves are only 1,778 PWh which would last about 37 years at the current annual consumption. All known resources (1,391 PWh) add another 32 years. Only when unconventional oil like oil sand and deep sea oil fields are utilized – which of course will be much more expensive and/or adding environmental concern – could we substantially prolong the usage of this energy carrier. In Figure 2.5 we see the rise of consumption of oil over the last decades (black line) and the annual findings in past years as well as potential future findings (annual bars). Until the 1960s, much more oil was discovered each year compared to the annual production and consumption. But the world oil discovery rates have been declining since the early 1960s and many of the large oil fields are now 40 to 50 years old. Moreover, we are now consuming oil at a rate about

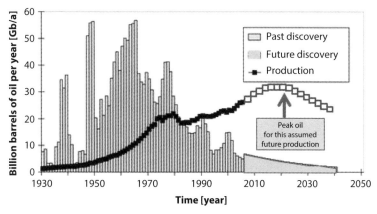

Figure 2.5 Annual regular conventional oil discovery and consumption (past discovery based on ExxonMobil (2002), revisions backdated).

6 times greater than the rate of discovery. There is one obvious trend: a growing gap is developing between the annual production and the future findings which lead to the fact that when the surplus of findings until 1980 (the crossing point between discovery and production) is eaten up, we will have reached the so-called point in time of "peak oil", also called "Hubbert peak". This means that from that year onwards there will be less oil produced and consumed than in previous years. In Figure 2.5 this is indicated for an assumed future production and happens in around 2020. Many projections conclude that this may already just have happened or will happen at the latest by 2040. This will undoubtedly have a pronounced effect on the price of oil within a free market dominated economy: whenever a product is known to be getting scarcer, it will experience an increase in price. This is quite different from the situation which we saw after 2008, when due to the financial crisis there was a lot of speculation which drove the barrel of oil towards $150. After that, it went back to the low $50s, increasing again towards and above the $100 rate by now. After peak oil has happened, there will only be one direction for the oil price without taking speculation into account, namely upwards towards $200 and even beyond. Many people see the risk of military conflict between countries demanding oil for their respective welfare.

For gas, we have some more reserves than for oil, but the peaking of gas will also occur in a few decades, depending on how much a further increase in findings will be. From all the different numbers I have seen, we may experience peak gas around the middle of this century.

Even if new findings – like shale gas in a number of countries or deep water oil drilling – are made, we should carefully balance the additional volume versus the ever increasing risks associated with the exploration of these additional resources. We all remember the catastrophe with BP's Deep Water Horizon on 20th April, 2010, where about 800 million liters of oil spilled into the sea. Imagine this had happened in the Arctic Ocean, where low temperatures would not have allowed to restore the environment with the help of bacteria as quickly as was the case in the Gulf of Mexico. In the case of shale gas, there are several environmental challenges associated with it which have to be examined very carefully. Let me just name one example where more transparency is definitely needed: even today (2013), companies which perform the fracking are in some countries allowed to refuse disclosure of the chemicals used for this process; in spite of the fact that one of the challenges is the contamination of ground water.

Coal is expected to be available for many centuries and even for 2 millennia with all known resources – if we follow the same annual usage as today. Unfortunately coal is adding most of the dangerous green-house gas CO_2 as will be discussed in more detail in the next section. Another important aspect when discussing the reach of fossil fuels are the underlying assumptions. Most often it is assumed that current consumption will remain constant in the coming years. However, if an increase in consumption is considered, many approaches based on stable consumption would prove totally wrong. Not introducing additional measures to replace the traditional energy forms would then have a catastrophic effect. The impact of a growing energy consumption can be impressively understood by a simple calculation done by Carlson [2-8]. He considered the US coal reserves to last for about 250 years at today's usage level. This time shrinks considerably to only half of that time if a very modest increase per year of just 1.1% is assumed. The duration further decreases to about 80 years at an annual increase of 2% in consumption. Qualitatively the same effect of considerable shortening is true for all of the worldwide fossil energy resources. In addition, if the gas and oil resources are depleted in a non-renewable economy, coal will have to be transformed into gas and liquid fuels.

2.4.2 Climate Change Due to Green House Gases – Best Understood by a Journey Through Our Earth's History From Its Origin Until Today

It is interesting to observe the different arguments related to the topic of increasing CO_2 concentration in the atmosphere and its influence on the global climate. While there are still some people who deny this correlation, there is an overwhelming number of serious scientists worldwide arguing in favor of it. The first group with the name "Climategate" is collecting scientific meteorological papers and looking for evidence that the topic of CO_2 is less of a problem and that it could even be viewed as an advantage. The second group, the Intergovernmental Panel on Climate Change (IPCC) brings together thousands of scientists from 194 member countries world-wide. It was founded by the World Meteorological Organization (WMO) and the United Nations Environment Program (UNEP) in 1988. The First Assessment Report (FAR) was published in 1990 and has been updated regularly since then (2nd AR or SAR, 3rd AR3 and 4th AR4 in 1995, 2001 and 2007, respectively [2-9]). The 5th AR5 report is just being prepared and will be ready in 2013/14. The findings of these reports are intensively discussed at regular conferences.

While it may be possible that in such a worldwide effort mistakes do also occur, every effort should be undertaken to avoid mistakes and to clarify the problems. Following a dispute with people from "Climategate" I tried to find out what the controversy was all about. As often when groups with different beliefs are arguing, it is valuable to first look for the data – which took me several clicks on the internet. One example of a mistake is the underestimated growth by the IPCC TAR3/AR4 [2-9] report of the Antarctic ice shield, while emphasis is only placed on the decrease of the Arctic ice. Let me just summarize a few numbers to illustrate the difficult situation between the opponents of the IPCC results, synonym "Climategate" and the IPCC community, synonym "IPCC":

1. From IPCC AR4 report: the change in the level of Antarctic sea ice was reported to amount to (0.47 +/- 0.8)% per decade
2. From Comiso&Nishio [2-10] a much higher number of (1.7 +/- 0.3)% per decade was reported as pointed out by "Climategate"
3. Newest data from Climategate (Turner, Comiso *et al* [2-11]) concluded 0.97 % per decade
4. Interestingly, in one of the "Climategate" comments there was also a cross reference to a discussion within IPCC discussing a number of (1.3 +/- 0.2)% per decade

Dated 16th February, 2010 the "Climategate" community summarized these results on the internet under the headline "Another IPCC Error: Antarctic sea ice increase underestimated by 50%". If the "Climategate" community knew that within IPCC the number as given in #4 above is seriously discussed they would have obvious difficulties with their headline (and then compare the number 1.3 from #4 with 0.97 in #3, which is even

30% higher!). On the other hand, there is obviously room for improvement as always in science – also in the IPCC community.

Another example from the "Climategate" community was an IPCC comment on the melting of the Himalaya glaciers. It became apparent that in an interview, a specific year, namely 2035 had been named as the year by which these glaciers may have disappeared. The IPCC later admitted that this had been a mistake. However, the same authors did agree that the melting of glaciers worldwide was a fact, but happening at a slower rate.

Regardless of when this will happen exactly, it will be a catastrophe for billions of people since glaciers are the prerequisite for a continuous flow of water to many of the rivers. No glaciers means no continuous water flow and as a result we will have times of no water in the rivers and in the same year flooding when the rainy season comes. This also has a dramatic effect on the situation of drinking water which is already a big problem in some regions today.

An important argument against the Climategate community is that the consequences of climate change are so disastrous that they would have to be 100% sure of the non-existence of climate change – which is impossible. Therefore they act irresponsibly.

A journey through time from our earth's origin until today
In a recent paperback by Rahmstorf and Schellnhuber [2-12] the topic "climate change" is described in a comprehensive and well documented manner. In particular, I would like to summarize one part where the authors took a journey through the earth's history from its beginnings until today. As this helped me to better understand what causes and consequences the various parameters have had on the climate of our planet, it may also be useful to others. I strongly advise everyone who is interested in the important debate on climate change to take a closer look to the mentioned paperback – it is more than worth it!

Before we start with our journey we have to understand what the major parameters are which determine our climate. At any given time, our climate is the simple energy balance between absorbed solar radiation (= incoming solar radiation minus reflection) and outgoing infrared radiation from our earth to the universe. Some gases, also called greenhouse gases, in particular CO_2, water vapor and methane let the incoming solar radiation pass, but not the outgoing infrared radiation, which originates from the absorption of the solar spectrum and change into a longer wavelength spectrum. The absorbed heat is equally radiated in all directions – therefore also a part back to earth. This results in a higher radiation on the surface of the earth (solar radiation plus reflected infrared radiation from greenhouse gases) compared with radiation in the absence of greenhouse gases. A new equilibrium with greenhouse gases can only be established if the earth's surface radiates more, with the consequence that it has to be warmer. It is this phenomenon which is called the greenhouse effect. This is also vital for life, as without the presence of these gases we would have a completely frozen planet: without greenhouse gases the average temperature on our planet would be -18°C and only with these gases is there a well-being temperature of ~+15°C. The relative proportion of the various greenhouse gases today is 55% for CO_2 and 45% for all others. The cause for concern is human's active contribution to considerably increasing the concentration of greenhouse gases, especially CO_2 in a relatively short period of time as we will see at the end of our time journey.

It all started 4.5 billion years ago when our solar system emerged from an interstellar matter on the brink of the Milky Way galaxy. The sun in the center is a fusion reactor where hydrogen atoms combine to helium and thereby release a lot of energy. Calculations show that at the beginning of our earth, the brightness of the sun was some 25% to 30% weaker than today. This should have resulted in a ~20°C colder environment on earth, well below freezing point. There would also have been more reflected solar radiation, also called Albedo, mainly because of the large area of ice, and there would have been less water vapor due to the lower temperature caused by the weaker solar radiation, so it is likely that the temperature would actually have been even lower. This should have resulted in a completely frozen planet over the first 3 billion years. However, there are many geological hints that during most of that time we had flowing water on our planet. This apparent contradiction is known as the "faint young sun paradox". To solve this contradiction, we have to assume that during the time of the weaker sun there had to be a higher greenhouse effect. But how is it possible that over billions of years there was always – or mostly – the correct mixture of greenhouse gases present to balance the changing radiation from the sun? The answer lies in a number of closed loop controls. The most important one is the carbon cycle, which has regulated the concentration of CO_2 over millions of years. Due to the weathering of stone – mainly in the mountains – CO_2 plus water reacted with many minerals[1]. Without a reaction in the opposite direction, CO_2 would have been disappearing from the atmosphere over millions of years, causing a significant temperature drop. Fortunately, the continents were drifting and as a consequence sediments were pushed into the magma and CO_2 was released into the atmosphere via volcanoes. Since the weathering depended on

---

[1] As an example, Olivin (Mg_2SiO_4), present in many volcanic rocks, was transformed into quartz ($Mg_2SiO_4 + 2\,H_2O + 4\,CO_2 \rightarrow 2\,Mg(HCO_3)_2 + SiO_2$).

climate, there was a closed loop: if the climate warms up, the weathering process speeds up, leading to more CO_2 being removed from the atmosphere, which decreases the greenhouse effect and counteracts a further temperature increase. However, this loop cannot cushion a quick change in temperature – the time needed for an exchange of CO_2 between the earth's crust and the atmosphere is much too long.

Several times, most recently ~600 million years ago, our earth experienced a so-called "snowball earth". This means that all continents, even the tropical regions, and the oceans were covered by an ice crust several 100m thick. The closed CO_2 loop helped overcome this deep frozen status: while the CO_2 reduction – which is the weathering process – stopped below the ice, the CO_2-source – volcanism – was still on-going. Over millions of years, the concentration of CO_2 increased steadily and even with the high Albedo the greenhouse effect became strong enough to melt the ice. The required concentration of CO_2 caused the temperature to rise up to levels of ~50°C and it took a long time to come back to "normality". Geological data give evidence that snowball earth periods were followed by times of high temperatures. Some biologists relate this most recent climate catastrophe as the cause for the evolution that followed.

Let us now take a closer look at the last 500 million years, for which we know the position of the continents and oceans and are able to construct the ups and downs of our climate by analyzing sediments. Two cycles can be identified: the first from 600 to 300 million years and the second lasting until today. The first cycle started with a concentration of ~5,000 ppm CO_2, which steadily declined to levels similar to those of today (~300 ppm). During the following ~100 million years it increased again towards 2,000 ppm and then continuously decreased again towards a range of 200 and 300 ppm. Those times with high levels of CO_2 correlate with little and mostly no ice coverage, while the times of low CO_2 levels show considerable ice coverage where the ice shields expand from the poles towards the 30th degrees of latitude. One piece of evidence proving the second warm phase without any ice coverage is the Cretaceous period from 140 to 65 million years ago. In this era, Dinosaurs even lived in polar regions, as evidence from archaeological findings in Alaska and Spitsbergen shows. There is one special finding worth mentioning as it is linked to what we are discussing and with our anthropogenic CO_2 addition. The above mentioned continuous cooling which started ~200 million years ago did not happen undisturbed: data obtained from the resolution of sediment from 55 million years ago show that in a time frame of less than 1,000 years there was a sudden temperature increase of about 6°C, which then took ~200,000 years to go back down to previous levels. The temperature increase can be attributed to an increase of carbon. Three possible causes are

- A release of methane from deep sea methane-ice
- The eruption of a huge volcano
- A meteor strike

A fourth possibility was elaborated by DeConto *et al.* [2-13]. These authors concluded that a thawing of the permafrost was a possible reason for this event, which would mirror what is happening just today. Whatever the exact reason was it can be seen that if a quick increase of carbon – like what mankind is doing now – takes place, this is quickly followed by a shift in temperature which then needs a long time to recover again.

We will conduct a more detailed analysis to the last 350,000 years, when the first hominids, including Neanderthals and Homo Sapiens appeared. From that time onwards, relevant data can be derived from the analysis of ice cores. The Wostok-ice core drilled in the 1980/90s in the Antarctic was well known, during which the CO_2 concentration (among many other data) over a time span of 420,000 years could be determined. The emerging picture from a multitude of such experiments is the clear evidence of three distinct cycles which can be summarized as shown in Figure 2.6.

The correlation of increasing CO_2 concentration with temperature increase and vice versa is obvious. An increase of CO_2 from ~200 ppm to ~290 ppm causes the temperature to rise by ~11°C. The same pattern

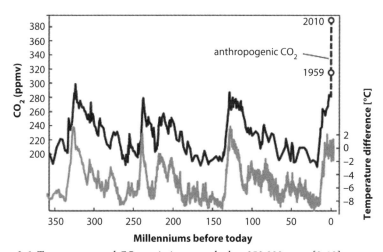

Figure 2.6 Temperature and CO2 variation over the last 350,000 years [2-12].

can be observed from the analysis of many ice cores with a calculated average temperature difference of ~8°C. This cycle happened three times and can be clearly associated to three ice-age cycles. It is interesting to note that the warming up only takes ~10,000 years and the cooling down is much longer, at ~90,000 years. The reasons for the ice-cycles are well understood today and can be contributed to a superposition of periodic ups and downs (Milankovich cycles [2-14]). It was possible to correlate these periodic changes and some extraterrestrial parameters: eccentricity of the earth's orbit (~ 100.000 years), the obliquity of the tilt variation of the rotational axis (~41.000 years) and the precession of equinoxes of the rotational axis (~23.000 years). Even if there are some critical discussions on the long 100.000 year cycle, the fact that there were causes of this cycle is well acknowledged. For example, Muller *et al.* [2-15] discussed as another possibility the periodic movement of the earth out of the plane defined by its way around the sun.

The cause and its effect are the exact opposite of today's situation. While today the increase of anthropogenic CO_2 causes the temperature to rise, in the past it was he different distance between the earth and the sun that caused temperature changes. As a result of the different temperatures, the CO_2 concentration in the atmosphere developed from the temperature dependent equilibrium between air and ocean (and some other effects). These long term changes will also happen in the future causing similar effects but should not be misused as an argument for continuing with business as usual.

The closer in time we get to today, the more detailed the available data become, to reconstruct even at a local level. Scientists are now able to locate about 20 incidents (the so-called Dansgaard-Oeschger incidents) with sudden and dramatic climate changes that took place since the end of the last ice-age. In Greenland for instance, there was a temperature rise of up to 12°C within only 10-20 years, which then remained for several centuries. Reasons for this were sudden changes of ocean currents in the North Atlantic, bringing huge amounts of warm water to this region. Also very large icebergs might have induced such changes.

While CO_2 concentration and temperature have remained fairly constant over the last 10,000 years, we can clearly see the distinct and quick increase of CO_2 produced by mankind over the last 150 years. We have to go back a couple of million years in earth's history before finding a concentration of the 390 ppm CO_2 that were measured in 2010 again. At that time in history, the ice shields were non-existent and this is exactly what has been happening over the last years: the melting of shelf ice in the Arctic and the disappearance of glaciers globally. Rising temperatures are additionally causing the Arctic permafrost to thaw, leading to a release of methane and CO_2 into the atmosphere. This will result in an additional temperature increase which will accelerate the above described developments. Schaefer *et al.* [2-16] estimated a decrease in permafrost area of between 29% and 59% and a 53–97 cm increase in active layer thickness by 2200. The corresponding cumulative carbon flux into the atmosphere has been calculated at 190 +/- 64 Gt Carbon. When comparing this with the current annual release of CO_2 which is ~33 Gt CO_2 (or ~9 Gt Carbon) we can see the significant amount which with high probability will come from thawing permafrost. Even on a short time scale the authors predict that from the mid-2020s onwards this permafrost carbon feedback will transform the Arctic from a carbon sink to a carbon source. In order to limit the temperature increase caused by green-house gases to the desired maximum of 2°C, an even larger reduction in fossil fuel emission is needed than is commonly discussed in the public debates.

According to the authors of [2-12] the changes in solar activity are only causing minor effects which can nonetheless be observed within the shorter timescale of the last centuries. By combining the knowledge of the past and by carefully analyzing the current trends, the authors discussed the following potential impact on our future climate:

- doubling of the pre-industrial level of CO_2 from ~280 ppm to 560 ppm would cause a temperature rise of (3+/-1)°C
- in order to fix the increase at only 2°C this concentration should not exceed 450 ppm which would correspond to a further addition of only ~750 Gt CO_2
- as we have already reached 390 ppm and are adding ~1.4 ppm every year there are only ~40 years left to reach a time with no further carbon release; even if we were to completely stop the burning of ALL fossil sources (NO gas, oil and coal) by then (~2050), we would still have methane released by animals, CO_2 from cement and permafrost thawing and much more – hence we will most probably be heading towards 3+°C in the future

This increased level of carbon dioxide and the accompanying temperature rise will have the following consequences:

- more and more severe weather extremes (hurricanes, tornados, cyclones, floods and droughts) – it is less the frequency of occurrence which changes, but the intensity and virulence

- global decrease in glaciers
- decrease of the area of the Arctic sea ice (an ice-free polar ocean during the summer is becoming reality)
- increased thawing of permafrost with considerable release of carbon dioxide leading to an even greater temperature increase
- rise of the mean sea level: cm to dm levels due to an increased volume of water caused by temperature increase, dm to m levels due to the thawing of existing ice shields (smaller levels for sea ice, +7 m if all ice shields in Greenland were to melt, +6 m if all ice shields in the West Antarctic were to melt and +50 m if the most stable east Antarctic ice shield were to melt). If all ice shields disappeared, as was the case in times of very high CO_2 concentrations tens of millions of years ago, there would be a sea level rise of more than 60(!) m – even 1m of sea level rise would have a catastrophic effect globally on areas inhabited by humans!
- Changes of ocean currents causing dramatic local climate changes

In summary, there is much more scientific evidence that supports a serious impact of anthropogenic carbon dioxide on climate change compared to the assumption that we may continue with business as usual. This was impressively demonstrated by two studies in 2004: one analyzed ~1,000 scientific publications to the topic "global climate change" [2-17] and the other ~600 articles in daily newspapers in the US (performed by the University of California). The scientific community supported to 75% the fact that man-made CO_2 and other green-house gases were indeed responsible to climate change; 25% did not comment to this specific point. The authors concluded that obviously among serious scientists there was a general agreement on anthropogenic CO_2 seriously affecting our climate. In contrast, the analysis of the newspaper articles showed that 35% of them emphasized the anthropogenic fact but also reported on the opposite, only 27% agreed entirely, while another 27% were of the opposite position. As the public debate is mostly influenced by the discussions in newspapers, we are unfortunately faced with a "balance as bias", as it was called. It is no surprise that the public debate is also heavily sponsored by all those think tanks which receive heavy financing by those industries benefitting from the "business as usual" approach and from denying the impact anthropogenic CO_2 is having on climate change.

2.5   Problems with Nuclear Energy

In the 1960s there was a big worldwide hype affecting all political parties and spurring them on to introduce nuclear power, to solve – which was the belief those days – all future energy needs: first with fission reactors, then with fast breeders and ultimately with fusion reactors. Although it was argued that this technology would be for peaceful application, we know that the cold war was also responsible for a lot of nuclear weapons. Additionally, I remember serious publications from these days arguing that by the turn of the century – which came and went 13 years ago – it would no longer be necessary to have electrical meters in the households because electricity from nuclear would be so cheap that it would be a waste of money to install such meters. Others argued that this cheap nuclear electricity could be used to keep the motorways free of ice in winter. The reality now looks quite different!

We know that investments in nuclear reactors are getting more and more expensive – especially if all possible accidents are to be taken into account – and also the uranium that is required is limited in quantity to last for some decades if only fission reactors are considered. This fact was and still is the reason for the need to develop fast breeder reactors which could indeed breed enough fissionable uranium and plutonium for the coming centuries as seen in Figure 2.4 and Table 2.6.

But this is no solution for the future – not in the short term and certainly not in the long term! Firstly, the Fukushima catastrophe in Japan in 2011 demonstrated – once again after Chernobyl in 1986 – that the residual risk should not be tolerated by society. It is frightening to see that only some months after the tragedy, when newspapers were full of articles, the topic has almost disappeared from the public debate by now. It is often discussed that in other countries such an earthquake and tsunami at the same time could not happen. But it is also known that terrorist attacks could also cause a melting of the core – anywhere and at any time! There is also a common misunderstanding of what the residual risk, specified in "accident per nuclear reactor years", really means. In 1993, a final study [2-18] was conducted by GRS (Gesellschaft für Anlagen- und Reaktorsicherheit = society for equipment and reactor safety) who came to the conclusion that every 20,000 reactor years such a tragedy may happen. Many people confuse this and believe that such an accident may happen in 20,000 years, by which time we will have solved the problems and therefore we should continue. But the term "reactor years" implies that we have to divide the 20,000 reactor years by the total number of reactors. As we have approximately 400 reactors running, the result is that we should expect such a tragedy every 50 years – and Chernobyl happened 25 years ago! Hence the result of the study was in reality 100% over-optimistic. In addition, the study concluded that with known safety measures this risk can be reduced by a factor of 10. If we take into account the "near major accidents" which happened 1979 in Three Mile Island (US) and 2006 at Forsmark (Sweden), it is obvious that we must conclude that there is a rather short term (~decade) real and likely risk of such an accident and not the nice-looking number of 500 years. There is no way to argue that by adding technical features this risk could be brought to zero – there is always a very likely probability that a disaster will happen. And imagine what would have happened in Fukushima if the wind had not been so gracious as to blow the nuclear cloud towards the ocean but would have instead steered it towards the Greater Tokyo area where 30 million people live – we would have witnessed an inferno.

Even if the above arguments were not convincing enough, there is another unresolved issue, namely the storage of waste from nuclear reactors. Nowhere in the world is there a viable solution. Even the storage in salt domes that is being considered in Germany can be questioned in view of the disaster of water penetration into the existing "Asse" salt dome in northern Germany. Knowing that we have to safely store the waste for many 100,000s of years, it is difficult to believe that any solution will ever be found. Researchers at the "Karlsruhe Institute of Technology" are experimenting on a method of converting the problematic long living radionuclides into much shorter living elements. This process is called "Partitioning and Transmutation (P&T)", where plutonium and the so called minor actinides (mainly Neptunium, Americium and Curium) are extracted chemically and thereafter transformed with energy rich neutrons into short living substances. This process would reduce the amount of high radioactive waste by a factor of 5, but would increase the quantity of low and medium radioactive waste several times [2-19]. Even the authors from the institute conclude at the end of their article that the ambitious targets set for nuclear reactors of the 4th generation – more security, increased sustainability, and higher profitability – can only be reached in combination with expanded closed nuclear fuel cycles, of which P&T would then be a natural part. Knowing that this process has been investigated since the 1960s without success, I very much doubt that this new nuclear alchemy would really be cost effective overall – if at all technically feasible.

Let us now take a look at what is often called the "third generation" of nuclear reactors, the "*fast breeder*". Without going into details there are two major arguments why I would not like this technology to be introduced. Firstly, the reprocessing of spent fuel is a must in order to extract the bred plutonium used in the fuel rods for further use. This implies that we would have to handle many tons of plutonium with the risk of proliferation. Secondly, and even more importantly, there is the fact that while it may be possible to construct an inherently safe fission reactor, to my knowledge this is not possible with a fast breeder. With small pebble-bed fission reactors and using ThO_2 as fission material the negative temperature coefficient after complete breakdown of cooling would prohibit a melting of the reactor core and thereby provide inherent safety. However, fast breeder reactors have to be controlled at every moment which is, as always in technology operated by mankind, simply impossible.

Finally a word on the development of *fusion* which looks very promising at first glance. It involves mimicking the reactions happening in the sun and with a few materials, we could have energy forever. But the challenges to be solved are enormous. When as a student in the 1960s I was listening to the lectures of Prof. Pinkau from the Max Plank institute in Garching (Munich), one of the advanced and knowledgeable places for this research, he stated that by the turn of the century a first pilot reactor should be up and running. When I discussed this with his successor Prof. Breadshaw from the same institute in the late 1990s he – and also his colleagues worldwide – were able to possibly envisage this pilot reactor in another 20 to 30 years from that time. For physical reasons, such a reactor should have a minimum size of 5+GW because of volume to surface considerations due to the high temperatures needed in the plasma, and should run only as a base load provider. Whether this is an appropriate solution for the future is very questionable, as decentralized power supply will be superior to an extremely concentrated and centralized supply. But even if all technical problems were to be solved, it is very difficult to see whether it will ever be possible to produce electricity below $ct5/kWh in the future with this technology when all cost elements, not only safety but also insurance, are taken into account – which is possible with many renewable technologies, including PV, as will be demonstrated later.

Looking to the European and worldwide spending on R&D for fusion one could critically question the huge amount of money already spent and even more importantly the money that will be spent in the future. In Europe, it was recently suggested that in order to complete the next pilot reactor ITER (International Thermonuclear Experimental Reactor in Cadarache, France) which needs more R&D money as originally scheduled, other R&D areas like renewables should decrease their R&D spending by the same amount in order to keep the total budget the same. As a physicist, I certainly support research in order to principally understand fusion – but for this we only need a small fraction of the money that would be needed to pilot a fusion reactor, which I deem to be neither necessary nor practically feasible.

I very much do like nuclear fusion – however, only when it is already up and running in our sun and the energy produced is transported as sunlight to earth where it can be cost effectively transformed into electricity and heat through photovoltaics and other renewable technologies.

3

The Importance of Energy Efficiency Measures

3.1 Traditional Extrapolation of Future Energy Demands or Alternatively "The Same or with Renewables Even Better Quality of Life with Much Less Energy"

There is a very simple calculation which is easy to remember and which demonstrates the false but often heard assumption that renewables are not able to power the future world energy demands. Using approximate numbers this goes as follows: Only one quarter of today's global population – about 1.5 billion in OECD countries – is using three quarters of today's primary energy – which is about 105 PWh. That also implies that today, three quarters of our global population only have one quarter of the primary energy left – clearly an unfair situation which has to change! As we would like to allow everyone in a future world to live in an environment with the same quality of life as we have, there is this simple calculation: At today's quality of life, 70 PWh of primary energy is needed per one billion people. If we assume that the global population will have risen to 10 billion people at the end of this century, this would quite simply mean that we would have a primary energy need of 700 PWh! This 5 fold increase in 90 years is often described in terms of a small annual increase, in this case only 1.8% per year, which seems little but underestimates the real challenges in the long term. It is often argued that such a huge energy content of 700 PWh cannot be delivered by renewable sources. This is wrong for three reasons: firstly, as will be explained later, when using renewables, we do not have to take the losses of converting primary energy – which is contained in solar energy, wind, geothermal and many more forms of energy – to secondary energy into account, secondly, it simply neglects the huge potential of intro-ducing energy efficiency measures with new technologies and last but not least even the 700 PWh could well be delivered by renewable sources only.

In this context it is wrong to talk about "energy saving" as many people associate this with meaning less energy for them and therefore a decrease in the comfort and convenience of their desired services, which they do not want. Instead, we must tell people that with new and energy efficient tech-nologies there will be "the same – or with renewables even better – quality of life but using much less energy". People must be trained to understand that this could imply that the initial investment in such new products may be somewhat higher, but that over a lifetime, the overall (sometimes called levelized) cost (investment plus running cost plus end of life cost) can be considerably lower than for established products.

In the following, we will analyze a few examples of such new technol-ogies which will be able to deliver the same service but with much less energy. As elements responsible for the decrease in losses as indicated by arrow (1) in Figure 3.1 we have among others: individual transport via

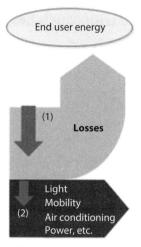

ref.: World Energy Council, Green peace, IEA,
own estimates

Figure 3.1 Energy efficiency measures to decrease the losses from secondary to end energy (1) and also today's need for end energy (2).

electric vehicles, lighting through solid state LEDs and OLEDs both in the industrialized world and in developing countries, and replacing pumps with inefficient part load characteristics with phase angle modulated power both for private and industrial use. Another important decrease in end energy as indicated by arrow (2) in Figure 3.1 is of a passive nature, for example through proper insulation of future houses. Obviously such houses no longer need as much energy for heating and cooling and in the future they will even need no or only little secondary energy compared to today's standard. Other contributions could come from better urban planning which should result in less transportation miles per inhabitant.

3.2   Decrease in End Energy Needs with a "Better Quality of Life"

3.2.1   Future Lighting: Energy Saving and Better Service

Today, lighting consumes a considerable fraction of global electricity. According to a publication by E. Mills from the Lawrence Berkeley National Laboratory [3-1] about 14% of global electricity or about 2 PWh is consumed through lighting. The break-down into the various consumption sectors was estimated as being 28% for residential, 48% for service, 16% for industrial use and 8% for street lighting and other applications. It should be remembered that this does not include the fuel- and candle-based lighting for the more than 2 billion people in developing countries. When taking into account the global average of electricity production from primary energy, lighting consumes about 6 PWh of primary energy. There is still a huge number of traditional light bulbs in use, which produce the needed lumens from electricity very inefficiently. Only about 10% of the electricity consumed is converted into light and 90% is lost as heat. Fluorescent lights are already better and have an efficiency of about 30 %. The advent of solid state lighting will change this picture considerably. Light Emitting Diodes (LED's) are already used in many applications as a bright point light source with a high efficiency of about 50%. The brightness has already reached the level needed in order to be used as head lights on upper class cars. The necessary production equipment and the required materials have now reached a mature state and will develop further in the future. Research is underway to explore laser based head lights with even higher efficiencies (the brand-new BMW i8 is expected to be equipped with laser head lights). Light tiles based on Organic Light Emitting Diodes (OLED's) which emit light homogeneously are now at the beginning of commercialization. These light sources emit light not from a filament or an arc, but homogeneously over an area. This is similar to a solar module, where light is captured over the module area and electricity is generated but works the other way round: electricity is fed into an OLED device and light is emitted all across the area homogeneously. Knowing that one can change the emission spectrum of the OLED device, it is possible to create very different colors, opening up a multitude of additional possibilities for designers, which is very important to create a new lifestyle product which will be attractive to many people.

Let us now draw two important lessons with respect to energy efficiency and cost from the example of lighting. The easy part is to understand that to create the same illumination in lumen (= light intensity), LED/OLED require a five times less energy than today's light bulbs. Hence, the secondary energy needed for the same quality of illumination is five times less than what it is today. Now, the more difficult part is understanding the "total cost of ownership". Let us take a simple example: a 60 W light bulb costs about one Euro and makes about 800 lumen over a lifetime of 1,000 hours. At an electricity price of 20 €ct/kWh the total cost when using the light bulb is 1€ + 60W × 1,000 hr × 0.2 €/kWh = 13 € (plus the cost for replacing the lamp every 1000 hours). An LED lamp costs much more for the same 800 lumen, e.g. 40 € but has a lifetime of 25,000 hours. In addition, it consumes less electricity. So the total cost of ownership here is 40 € + 12 W × 25,000 hr × 0.2 €/kWh = 100 €. For the same 25,000 hours, the total cost of ownership for the light bulb would be 25 times the above mentioned 13 €, which is 325 €,; more than three times more expensive. Hence, the new LED lamp is – although more expensive at first sight – less expensive for the same service. It becomes more and more important – and this is also true for many other new technologies – to take an integrated look at the whole lifetime of alternatives before making a decision. This is especially important for renewable technologies producing electricity when compared to traditional fossil and nuclear ones: while the initial investment for renewables is considerably more expensive, over their whole lifetime the cost of fuel is zero in the form of wind and solar. Hence the so called "levelized cost of service (LCOS)" which includes all cost components over a lifetime becomes the important number to look at when comparing. In our case, the service is lumen multiplied by time. For the traditional light bulb, we obtain a cost of 1.6 €ct/(1,000 lm hr) and for the LED lamp the cost is only 0.5 €ct/ (1,000 lm hr). It is also important to clearly specify whether the stated levelized cost is "real LCOS" or "nominal LCOS". In the example above, we have neglected inflation over time which gives us the "real LCOS" (by definition). If we had included inflation (e.g. 2.5%), the resulting "nominal LCOS" would have increased by 10% to 20%, depending on the (lm hr) per year. Whenever numbers for LCOS are compared, it is important to state clearly which of the two possibilities has been chosen in the calculation.

Taking a global look at what can be saved in electricity whilst maintaining the same quality of illumination, it can easily be seen that replacing today's portfolio of light bulbs and other light sources with solid state LED's and OLED's could decrease the 2 PWh electricity need to only 500 TWh.

3.2.2   Electro-Mobility: Powerful and Halving Consumption (But Only If Electricity Comes From Renewables)

It is interesting to note that when the automobile was first developed in the late 19th century, the first thing to be considered was an electric drive. Due to the fact that with oil it was more convenient, everyone has concentrated on the Otto and diesel engine until now. The advantages of electric drive compared to traditional engines are: a much easier and smoother acceleration due to the torque dependency on the number of revolutions, which is much better than in a traditional engine. We all know this from by the experience of riding a tram. In addition, one does not need all the mechanical devices like clutches, gear boxes and ultimately the motor is part of the wheel, which will give even more freedom to the designers of future cars. A very important feature of electric vehicles is the much higher efficiency when comparing the secondary energy input. Electric motors have an efficiency of more than 90%, the battery efficiency is also in the range of about 90% (or better in the future) and all other processes to enable the movement of the car have another +/-70%. All together we have an efficiency of $0.9 \times 0.9 \times 0.7 = 0.57$ or about 60%. This can be compared to a diesel engine which delivers only up to 30% of the secondary energy diesel fuel to the motion of the car. It is important to note at this point that there has to be an important prerequisite to really make the electric car superior to the traditional diesel car: namely, we have to compare the primary energy inputs in our traditional world. For the electric car this means that on a global level for every unit of electricity, two units are lost. This would make the electric car only 20% efficient, which is less efficient than a state of the art diesel engine. Only if the electricity is produced by renewables do we have the true benefit of electric cars which are more than twice as efficient as traditional cars.

It should also be noted that today most transportation for private cars, trucks, buses, ships and planes is powered by petrol – with the exception of railways which run on electricity in many countries. This will change in the future and as in the case of long range trucks, ships and planes electricity will not be the fuel of choice. Instead, these vehicles will be powered in the future either through biofuel or "power to gas". The latter is obtained through hydrogen produced by water electrolysis (with renewable electricity) which reacts with CO_2 to get CH_4. In the case of trucks and ships I could imagine the use of hydrogen (or gas with a reformer, which splits hydrogen from the used gas), if an infrastructure can be envisaged. For planes, although it has already been demonstrated in the past that jet engines could use hydrogen, this may not be a good solution as the exhaust fumes will emit water vapor. This is more dangerous in the stratosphere as the small H_2O water vapor molecule acts like CO_2 but stays in the atmosphere much longer.

For the majority of private cars, it is well known that most of the cars drive less than 100 km per day, which is a distance easily covered even by today's electric cars. I can see an easy adoption in areas where many families have a second car which is often hardly used for distances above this range. Even when the first car travels less than the mentioned 100 km per day on most days, it is desirable to have the possibility of driving longer distances on some days. This could be done in a variety of ways: either one could have a car-sharing program for such rides, or one could use cars which are already available today from Opel (the Ampera) which have a small 1.4 l / 54 kW engine only used to charge the batteries and making all the positive advantages of the electric drive possible. Since November 2013 BMW also offers such a range extender with its all-electric i3 car. Alternatively, one could also use hydrogen either in a combustion machine or in combination with a small fuel cell producing electricity to recharge the battery. Other possibilities could combine the use of long distance trains and renting electric cars at the point of destination.

In summary, one could envisage the following situation with future transportation:

- **Private cars:**
  most with electrical engines, many with an additional range extender (turbine with fuel/power gas or fuel cell with hydrogen)
- **Public transport by buses:**
  local range buses in the urban sector could efficiently use electric engines; for long range buses one could also use a range extender as with private cars
- **Trucks:**
  in many cases one would use biofuel to be most flexible; electric drive with range extenders is also a possibility
- **Ships**
  could be run either by fuel cells with hydrogen or on biofuel

- **Planes**
  most probably running on biofuel

Summarizing the potential for the transportation sector impressively shows how we could decrease the energy needs but keep the same level of comfort for all transportation activities. It should also be highlighted that when we speak about hydrogen it only makes sense when the electricity for water electrolysis comes from renewable sources. The following picture could emerge for the secondary energy used:

- **Private cars**
  Assuming the majority (~3/4) of driving distance is covered with electricity and the remaining distance is covered with petrol/power gas or gas/H_2, approximately one third of today's secondary energy could be saved.
- **Buses**
  Assuming half of the cumulated distance traveled is in or near towns and the rest of travel is powered by petrol or gas/H_2, about one quarter could be saved.
- **Trucks** will see little and **ships and planes** will see no change.

With the split of energy used in the various transportation sectors described in the preceding chapter and the specific energy needs for the different applications, we can estimate the potential savings today if we assumed that future technology will be implemented. For the same quality of transport, this saving could be 40% or, in energy numbers today's 24 PWh could be reduced to 14.4 PWh.

3.2.3 Comfortable Houses: Properly Insulated, Facing South (In The Northern Hemisphere) and Producing More Energy than Needed

In many countries, a major part of primary energy is used to provide the heating and cooling of buildings. For the example of Germany, this amounts to approximately 40%. Hence, a relatively large amount of money is spent by individuals to buy gas and oil for heating and cooling purposes. Obviously, this is an area where we could drastically decrease the amount of energy used today.

Technologically, this topic is one of the easiest, but in reality it is one of the biggest challenges to overcome. It all starts with the proper orientation of the buildings – at least the future ones as quickly as possible – with the roof and the respective rooms facing south and north. If we go back to ancient times, people already knew how to make the best use of orientation in order to have cool houses in the southern regions and not to need too much heat for houses during winter in northern parts of the globe. Unfortunately, most of this knowledge is no longer used by most architects and urban district planners, but it could provide the basis for integrating the local situation of solar radiation to ensure the comfort of a house. The easiest way to make the best use of a roof for solar capturing is the orientation of houses with one roof facing south. All that needs to be done is a proper orientation of the roads and guidance on how to orient one roof towards the south. This was already done years ago in North Rhine-Westphalia in Germany, when the Green party pushed for integrating more renewables and implemented such a regulation for new district buildings. It would be a major achievement if we could all agree on such simple measures which do not cost additional money – they just have to be planned properly!

While new houses now have all the possibilities to include proper insulation at a reasonable price, the installation of the different electrical wiring and piping for cold and hot water is necessary, too. It remains to be seen when it makes sense to have a dual wiring for DC and AC in a house, knowing that many of the appliances of a house today use DC electrical power provided by transforming AC. This includes most of the audio and IT systems, solid state lighting and many more. Taking into consideration that a future house will have a DC PV system with a several kWh battery system, one could save energy which otherwise gets lost by transforming AC into DC, especially for the many small appliances. It also greatly enhances the self-consumption of electricity produced within the same house, at least within the low voltage region where the house is located. These considerations are very different for existing houses, which only undergo major renovations at a rate of about 1% to 2% per year, during which new state of the art features can be implemented at a reasonable price.

A good way to push for implementing state of the art technologies in new buildings is the new European directive which gradually gives a mandatory goal for proper insulation in the first place and goes further into a future house, which produces more energy than it consumes over the course of a year. This will consequently start with passive measures like insulation but will increasingly include the production of electricity and heat at the point of its use, which is perfectly well provided by decentralized PV systems and thermal collectors on the individual houses. It should also be highlighted that the insulation of houses alone is not enough for the well-being for the residents: in order to prohibit the growth of mold fungus if the house is not carefully ventilated (…and most people are just too lazy), a well-functioning forced ventilation must also be installed. It will still require a lot of work for a standard package at a reasonable price to be available for existing houses.

An interesting question arises in the longer term as to the provision of heat in a well-insulated house in the future. The residual heat for achieving the right temperature becomes very little and one could wonder whether the hot water that is needed could be provided more cheaply by individual electrical heating at the point of use, thereby eliminating the additional future piping for hot water. This discussion brings me to a more fundamental point. While it makes most sense today to use solar thermal systems and more and more district heating – as it has been nicely and successfully done in Denmark – integrating seasonal heat storage from summer to winter in large quantities will also determine the appropriate infrastructure of the whole district NOW, including the provision of hot water piping in all of the houses. Over the lifetime of such systems (…and beyond) this will determine the way heating will be done in such an area. I would argue that anything we can do today for the coming years by using renewables like solar thermal is OK, even if in future years there may be a more cost efficient solution available for new houses in new districts.

3.3   Today's Energy Needs with Known Energy Efficiency Measures

The potential split of sectors and electricity for today's world would look fundamentally different to Figure 2.3 and is shown in Figure 3.2. For the various sectors, I assumed a decrease in energy needs of ~-40% for mobility, ~-80% for low temperature heat, ~-20% for the rest of the industry and ~-40% for electricity in industry as well as for the rest of the electricity consumption. The original secondary fossil fuel for transportation is assumed to shift completely to electricity or hydrogen/fuel derived from renewable electricity (in the latter case we have to provide an additional fraction of electricity corresponding to the associated losses). The total secondary energy would shrink from 90 PWh to approximately half of that (47 PWh) – in other words we would have an increase of the energy efficiency by a factor of 2. Total electricity would account for 54% of secondary energy, low temperature heat would decrease to 11% and process heat for industry and SMEs would be at 35%.

Before we take a closer look to the potential and today's situation with the most important renewable energies, two different approaches for pushing energy efficient products as well as passive measures should be analyzed.

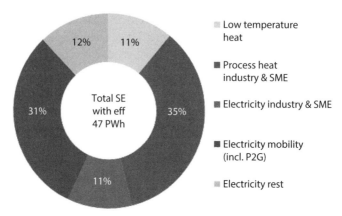

Figure 3.2 Today's secondary energy sectors with known conservative efficiency and passive measures, which would only be ~47 PWh compared to the actual 90 PWh.

3.4   Support Mechanisms to Facilitate New Products: Ban The Old or Facilitate The New Ones

Let us assume that after a thorough analysis society and politics came to the conclusion that a better product compared to an existing one should be used in the future, for example energy efficient light devices. As always in the beginning of the lifecycle of new products, the cost will be high which should result in a high price. There are two fundamentally different ways to facilitate the introduction of such new products: politics could ban the old product by law – which happened in Europe with the traditional light bulbs – and force the industry to produce new products which were not working as they should at the time the law was put in place. Solid state lighting devices where not yet being mass produced and in the case of the energy efficient bulbs, it was realized that they only work if poisonous mercury is used, causing major difficulties when disposing of them after use. So obviously the demon "high energy consumption" was replaced with the demon "highly poisonous substances". The big mistake in this approach was that there was no intensive discussion between industry and politics.

There is obviously a much better way, as demonstrated in Japan with the so called "Top Runner" project. In short it works as follows, using the hypothetical case of energy efficient light bulbs: on an annual basis the industry is asked to create the best efficient bulb for the existing technology status. With the help of a well thought out evaluation procedure with relevant figures of merit the best e.g. 3 products from companies A, B and

C are recognized as the best ones and receive a big push for free advertising and other promotional activities. The following year, only the three chosen products are allowed to be sold, all others are then banned and the same procedure would take place again and so on. Most probably we would have avoided the energy efficiency bulbs containing mercury and instead would have created an interim solution with bulbs containing halogen before an even more efficient solution in form of solid state LEDs and OLEDs in the future would have also penetrated the market.

As a general observation, I see major drawbacks to bureaucrats simply banning something without properly consulting with industry. It is much more advisable to reward the best product but to leave the details up to the respective industry. This definitely works for consumer products; strategic products which will be defined and discussed in Chapter 5.1.1 may need a different procedure.

4

Overview of the Most Important Renewable Energy Technologies

4.1 Basics About the Potential of Various Renewable Technologies

In the preceding chapters the finiteness of traditional exhaustible energy sources including their harm to the environment and mankind was discussed. If we just compare the reserves for the various exhaustible energies with the annual primary energy consumption, we can see in Figure 4.1 that with the exception of coal these reserves will only last for a few decades (see also [2-7]). However, if we compare this to the annual solar energy radiation we can easily see that even a substantial increase in exhaustible energies could never reach the input of the annual solar radiation. Taking a look at the various renewable technologies as seen in Figure 4.1 like wind, biomass, geothermal, wave and tidal as well as hydro energy, we realize when comparing their annual energy delivery to that of solar radiation on the continents that solar has by far the biggest contribution to offer [2-7]. It is therefore no surprise that solar will also have the most pronounced place within the portfolio of the future scenarios discussed in

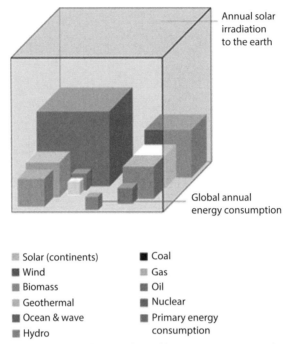

Solar (continents)   ▪ Coal
▪ Wind   Gas
▪ Biomass   ▪ Oil
Geothermal   ▪ Nuclear
▪ Ocean & wave   ▪ Primary energy
▪ Hydro   consumption

Figure 4.1 Worldwide energy offer for exhaustible primary energies and annual solar radiation in comparison to the annual primary energy needs and technical potential for the annual energy from various renewables (source EPIA).

later chapters. Figure 4.1 shows the total annual renewable Energy input for each energy form. The technical potential given in Figure 4.1 for the various renewables has also been analyzed by WBGU [2-7]. The numbers they arrived at for the technical potential can be seen in Table 4.1. In addition this group also worked on a so-called "sustainable potential" which takes into account important restrictions. These are described in more detail in [2-7] by going from the "technical potential" to the "sustainable potential" and can be highlighted as follows for the individual renewable resources:

- **Biomass** should not compete with food production and biodiversity should be preserved
- **Geothermal** for electricity production is restricted to those regions which supply high temperature steam
- **Hydro** should take into account the methane emissions from flooded areas, the siltation of large reservoirs and the impact on biodiversity

Table 4.1 Global potentials for renewable energy sources (for comparison: global primary (secondary) energy consumption in 2010 ~140 (90) PWh).

	Technical potential [PWh/year]	Sustainable potential [PWh/year]
Biomass	224	28.0
Geothermal	202	6.2
Hydro	45	3.4
Solar	78,400	2,800.0
Wind	476	280.0
total	79,347	3,117.6

Source: WBGU Flagship Report [2-7]

- **Wind** is restricted by noise emissions near to buildings for on-shore farms
- **Solar** has limitations when solar farms are competing with food production

As a reminder, the split between the major renewable sources for the sustainable potential can be summarized as follows: 90% solar, 9% wind and 1% all other renewables (including hydro). We will later discuss a model where for the future secondary energy needs the fraction of solar is reduced to only 60%, that of wind increased to 20% and that of all other renewables substantially increased also to 20%.

It is often argued that land use for renewables may not allow to generate enough energy. Figure 4.2 demonstrates that this argument is wrong. If we used solar modules as of today we could produce the annual electricity of the world (~20,000 TWh) or the EU (~3,000 TWh) as indicated by the two red areas. This already results from a simple calculation with conservative assumptions:

- 15% module efficiency (today even higher)
- With realistic solar insolation for the Sahara desert we have 1.8 kWh/$W_{PV\ installed}$
- Installation area = 2 × module area

The necessary area to cover the global electricity consumption can then be calculated as:

$$(20,000 \times 10^{12}\ \text{Wh} \times 2) / (150\ W_{PV\ inst}/\text{m}^2 \times 1.8\ \text{kWh}/W_{PV\ inst}) = 385 \times 385\ \text{km}^2$$

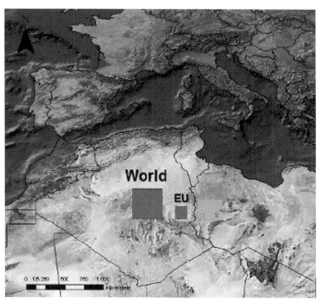

Figure 4.2 Area with PV modules needed to generate the world's and Europe's annual electricity needs.

Similarly, the area needed to cover the annual European electricity consumption would be 150×150 km².

When I am showing such a relatively small area in lectures and discussions I mostly encounter disbelief – but check it out yourself! This little calculation should only demonstrate the small area coverage needed for the global electricity needs. The nice thing is that in many cases we can use areas already in use where there are buildings, parking lots and many other areas that could be covered with PV modules.

Similar examples could also be given for other renewable technologies.

4.2   Wind Energy

There are many excellent books and journals describing this renewable Energy form. A good overview is given by Darrell Dodge [4-1] and up-to-date market and technology information can be researched on the website of the European Wind Energy Association EWEA [4-2].

The history of wind power dates back more than 1,000 years ago. A vertical axis system was developed in Persia about 500-900 A.D. We know that similar vertical axis windmills were developed in China with one piece of documentation dating back to 1219 A.D. although it is believed that the windmill was invented in China as long ago as 2,000 years. The windmills were most commonly used for water pumping and grain grinding. The first illustrations of European windmills on the Mediterranean coast (1270 A.D.) show a four bladed mill with a horizontal axis. This type has a higher structural efficiency compared to the vertical axis due to less shielding of the rotor collection area. In the late 14th century, the Dutch started to refine the tower mill design. An important improvement to the European mills was the introduction of sails. These could be formed to generate a higher aerodynamic lift and they provided increased rotor efficiency compared to the vertical axis mills. The next 500 years saw a continuous development and by the 19th century the windmill sails had all the main features that we see today in the modern wind turbine blades. Famous windmills are the thousands of sail-wing windmills on the Lassithi plateau of Crete, which were built in the late 1800s and are indispensable for local farmers to pump water with many of them still in use today. The development of the multi steel-bladed water pumping windmill was perfected in the US during the 19th century starting with four paddle-like wooden blades, followed by an increase in the number of thin wooden slats. Steel was used from 1870 onwards because it could be made lighter and worked into more efficient shapes. Between 1850 and 1970, over six million such small windmills were installed in the US alone.

The first large windmill to produce electricity was built as early as 1888 by Charles F. Brush in Cleveland, Ohio. The machine had a multiple-bladed rotor which was 17 meters in diameter and astonishingly operated 20 years long. The output was 12 kW which can be compared to the approximately 100 kW produced by today's machines of the same diameter. Only a few years later, in 1891, the Dane Poul La Cour developed the first electrical output wind machine incorporating aerodynamic design principles. The use of 25 kW windmills had spread throughout Denmark by the end of World War I but came to an end due to cheaper and larger fossil-fuel steam plants. Also worth mentioning is the development of modern vertical axis rotors in France by G.J.M. Darrieus in the 1920s, further development of which had to wait until the concept had been reinvented in the late 1960s by two Canadian researchers.

A first utility-scale wind energy conversion system was developed in Russia in 1931 with the 100 kW Balaclava windmill. This system operated for about two years on the shore of the Caspian Sea, generating 200,000 kWh of electricity. The largest system was built in the US and was a 1.25 MW machine, installed in 1941(!) in Vermont. After only a few hundred hours of operation, one of the blades broke off, apparently as a result of metal fatigue. Obviously, the jump in scale was too big for the materials available. German engineers should have learned from this finding 40 years later, when they decided to build the 3 MW "GROWIAN" wind mill.

After the oil crisis in 1973, a government funded program was started in the US to test a variety of different small, medium and large wind turbine designs. Up to 3.2 MW horizontal axis and several vertical axis Darrieus machines were tested. In Canada there was even the installation of a 4 MW Darrieus machine within the Hydro-Quebec project Eole. This wind turbine operated from 1988 to 1993, generating about 12,000 MWh electricity [4-3]. Also in Germany a big technology program was initiated with substantial support from the government, the Great Wind Mill (GROWIAN = GROsse WIndenergie ANlage) with 3 MW power which was installed in 1983. Similar to the problems in Vermont (see above), the up-scaling was too dramatic and the windmill was decommissioned in 1987 after only a few hundred operational hours.

It is interesting to analyze why in the 1980s it was not possible to translate this technology driven program in the US into a commercial success with GWs of installed windmills. The major reasons for this were: too great an interference of political factors, too early a withdrawal of financial support and – most important in my point of view - technology push rather than market pull support. We will also be looking at more examples for other renewable technologies.

Let us now take a look at which factors will drive the further technology development in the future. In Figure 4.3 the windmill picture on the right shows a typical windmill with a tower and the electric generator at the top together with three blades, defining the radius of the rotor. It is important to understand that the power output of a given windmill is proportional to the third (!) power of the wind velocity. This means that an increase by only 20% (30, 50 or 100%) will increase the power output by a factor of 1.7 (2.2, 3.4 or 8). This is important for two reasons: one is that at a given site the wind velocity increases from ground to higher levels as seen in the left hand graphic of Figure 4.3 – very quickly at sea or on large lakes and much less in forests or towns, where the landscape roughness at lower levels degrades the wind velocity. This is the reason why windmills get higher and higher with their hub height. The other reason is that the annual output of a windmill at a given site varies quite significantly from year to year as the average wind velocity changes. Another important factor which determines the power output is the radius of the blade (or diameter of the spinning rotor). This dependence is proportional to the square of the blade size.

The continuous development over the last 20 years driven by market pull programs like the feed-in tariff developed in Germany (and explained in more detail in the next chapter) is illustrated in Figure 4.4.

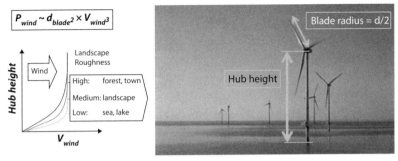

Figure 4.3 Principles for wind energy.

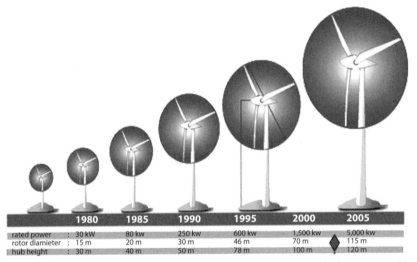

Figure 4.4 Continuous development of wind energy converters over the last 30 years (the blue diamond marks the power of the German GROWIAN 3 MW wind mill in 1983).

As can be seen nicely, the factor of up-scaling in terms of power output or rotor diameter is only ever in the range of two to three and not – like in the case of the GROWIAN an up-scaling factor of 100 (30 kW to 3 MW). Unsurprisingly we now have 5 MW windmills in operation and the further development towards 10 and even 15 MW per machine is underway by further increase of hub height and rotor diameter. Another impressive increase due to the above mentioned dependencies is the power output at a given site for a single windmill increasing from 35 MWh to 1,250 MWh and 17,000 MWh for the 30 kW, the 600 kW and the 5 MW machine, respectively. For each doubling of installed power capacity there is an increase by almost a factor of 3 in energy output.

For a new industry, it is important not only to develop the respective product – in our case the windmill – but additionally the infrastructure and

Figure 4.5 Auxiliary systems for installing a 5 MW wind energy system [RE Power].

auxiliary systems also need to become large scale. In the case of windmills, this can be impressively seen by looking at Figure 4.5, which shows the installation of a 5 MW machine. The large cranes are now available in sizes and quantities which make it possible to install many big windmills in a short period of time.

The next big challenge is the development for off-shore windmills. One of the driving factors is the high wind velocity as previously explained. But even more important is the fact that unlike on-shore, where the wind only blows 1,500 full load hours, this increases greatly with off-shore to 3,000 up to 4,000 full load hours. This becomes important in view of the future of our 100% renewable goal. But before this can be utilized we have to overcome some technical hurdles. Firstly, the harsh environment – waves, strong wind and corrosion due to salt – and again the infrastructure to transport the big components and install them with specially constructed ships.

But not only the building of GW off-shore wind parks is an important piece of additional work that needs to be done. All the TWh of electricity that are produced at an off-shore park then have to be transported to the millions of electricity users, which means electricity transportation firstly to land and then distribution via High Voltage grids to the metropolises for usage.

While the generation cost for large on-shore windmills today are in the range of 5 €ct/kWh, the cost is much higher for off-shore production. It is not known today what the real lifetime for off-shore windmills will be in the future, especially taking into account the harsh environment. In order to account for this uncertainty, the feed-in tariff paid in Germany for such projects is in the range of about 18 €ct/kWh – well above even small roof-top PV systems. This does not include the connection from the off-shore plant to the coast, which adds another one to two Euro cents per kWh.

How did I calculate this connection cost number? I listened a lecture by Prof. Breitner from the University of Hannover, in which he reported on three specific investment numbers for currently installed wind-parks in the North and Baltic Sea: for a 580 MW (85 km subsea cable), 1,200 MW (125 km subsea cable) and a 800 MW (75 km subsea cable) off-shore wind-park the investment by the company Tennet was €700 million, €1,200 million and €850 million, respectively. If we simplify this to make it easy to remember, we can say that in all three cases the connection cost was about €1 billion per GW or $€1/W_{wind\ power\ installed}$. Knowing that in Germany a PV system in 2011 had a generation cost of about 20 €ct/kWh at a 2.5 €/W installed system price, 20 years of depreciation and 1,000 full load hours, we can easily calculate the following:

$$20\ €ct/kWh\ /\ ((4{,}000/1{,}000)_{full\ load\ hours} \times (2.5/1)_{invest\ per\ W}) = (1-2)\ €ct/kWh$$

The result depends on whether we assume the same depreciation of 20 years or a 40 year depreciation for the electricity grid.

In Table 4.2, the cumulative installed wind power at the end of 2012 for the Top 10 countries together with their respective share as well as the ranking for the installations in the year 2012 are summarized [4-4].

When comparing the annual installations in 2006 and 2012, we see that the global installations went up by a factor of 3 (which corresponds to an annual growth rate of 20%). It is impressive to see how particularly China has increased its wind power installations. China went from No. 5 to No. 1 in only 6 years by increasing the installations by a factor of 10(!). In the US, the factor of increase is merely 5, Great Britain and Italy increased their installations by a factor of 3, India, Germany and Canada increased theirs only slightly by a factor 1 to 1.5 while in Spain a decrease of 20% can be observed. This overview of annual installations in the different countries is even better reflected in the cumulative installations. From 2006 until 2012, China went up from only 2.6GW to 75GW (almost a factor of 30), the US figures increased from 6.3GW to 60GW (a factor of ~10), France's numbers went from 1.6GW to 7.6GW (a factor of 5), cumulative installations in Italy and the UK only quadrupled, in Portugal and India it tripled, Spain doubled their cumulative installations, while Germany only increased

Table 4.2 Cumulative (2012) and annual wind power installations for the Top 10 countries in 2012 and 2006 for comparison.

Country	Cumulative end 2012 (in GW)	Share in % (Ranking in 2012)	Installations (in GW) in 2012	Share in % (Ranking in 2012)	Installations in GW and share (in %) in the year 2006 [ranking]
China	75.3	27 (1)	13.2	30 (1)	1.3 (9) [5]
USA	60.0	21 (2)	13.1	29 (2)	2.4 (16) [1]
Germany	31.3	11 (3)	2.4	5 (3)	2.2 (15) [2]
Spain	22.8	8 (4)	1.1	3 (7)	1.6 (10) [4]
India	18.4	7 (5)	2.3	5 (4)	1.8 (12) [3]
Italy	8.1	3 (7)	1.3	3 (6)	0.4 (3) [8]
France	7.6	3 (8)	–	–	0.8 (5) [6]
Great Britain	8.4	3 (6)	1.9	4 (5)	0.6 (4) [8]
Canada	6.2	2 (9)	0.9	2 (9)	0.8 (5) [7]
Portugal	4.5	2 (10)	–	–	–
Brazil	–	–	1.1	2 (8)	
Romania	–	–	0.9	2 (10)	
RoW	39.9	14	6.4	15	3.1 (21)
Total	282.6	100	44.8	100	15.2 (100)

them by a factor of 1.5. Denmark, which was not among the top 10 in 2006 and 2012 in terms of annual installations, is remarkable in that at the end of 2011 the cumulative installations in this rather small country contributed almost 26% to the annual electricity consumption in Denmark. This is the highest share of wind power penetration in one single country. Even if we add the installations in all EU countries together in 2012, we obtain about 8 GW [4-5]. This places the EU before the US but is still only 1/2 of the Chinese market. Only a few years ago, it was mainly European companies together with GE Wind Energy from the US that exported windmills to these countries. This has drastically changed in only a few years. The increased installations, especially in China, have been utilized for the creation of companies producing windmills including all necessary components (blades, transformers etc.) as seen in Table 4.3 [4-6,7].

Comparing the 2010 and 2005 data, only the market leader Vestas could maintain the pole position, although this was accompanied by bisecting its market share. Similarly, the big ones in 2005 from the US and Europe also bisected their market share (GE Wind energy, Enercon and Gamesa), only Siemens Wind Power and the Indish Suzlon Group could maintain their (small) market share. The Chinese manufacturers took advantage of their strongly increased home market by impressively increasing their capacities: Sinovel, the new No. 2, overtook GE Wind Power (which became No. 3) by increasing their capacity by a factor 15. Goldwind even increased their capacity 25 fold, making them No. 4. Dongfang and United Power from China were not visible in 2005 but could gain market positions 6 and 10, respectively. Other Chinese companies like Mingyang, Sewind and Hara XEMC, listed as No. 11, 14 and 15 in the world, respectively, are emerging to increasingly challenge the positions of the traditional market leaders from Europe and the US. In summary, in just five years the Chinese manufacturers increased their global market share from about only 3% in 2005 to an impressive 35% in 2010. The picture in 2012 changed due to a decrease in the Chinese market share within the Top 10 to 26%. This change was mainly caused by a market share doubling of the European company Siemens Wind Power and a significant drop in market share of the Chinese companies Sinovel and Goldwind.

The ever increasing appetite for energy in China paired with huge amounts of money to boost investments for capacity increase makes this country unique in the world. Many studies conclude that China will become the biggest economy after 2030. However, if the growth of the Chinese industry continues at a similar pace in the coming years with European and US growth being significantly lower and renewables being a fast growing sector worldwide, it will not take another 20 years and China could well

Table 4.3 Top 10 manufacturers in 2010 and 2012 together with comparison in 2005.

Company [ranking 2010]	Country	Market share (%) in 2010 (2012) [2012 ranking]	Sold GW in 2010 (2012)	For comparison: market share (%) in 2005 [ranking]
Vestas [1]	Denmark	14.8 (11.8) [1]	5.8 (5.7)	28 [1]
Sinovel [2]	China	11.1 (2.7) [9]	4.4 (1.3)	2 [7]
GE Wind Energy [3]	USA	9.6 (11.8) [2]	3.8 (5.7)	18 [2]
Goldwind [4]	China	9.5 (6.0) [7]	3.7 (2.9)	1 [8]
Enercon [5]	Germany	7.2 (7.2) [4]	2.8 (3.5)	14 [3]
Suzlon Group [6]	India	6.9 (6.6) [5]	2.7 (3.2)	6 [6]
Dongfang Electric [7]	China	6.7 (<2.3)) [–]	2.6 (<1.1)	–
Gamesa [8]	Spain	6.6 (6.4) [6]	2.6 (3.1)	13 [4]
Siemens Wind Power [9]	Germany	5.9 (11.0) [3]	2.3 (5.3)	6 [5]
United Power [10]	China	4.2 (3.5) [8]	1.6 (1.7)	–
Sewind [–]	China	– (2.3) [10]	– (1.1)	
RoW		19.5 (30.7)	7.7 (33.5)	12
Total		100 (100)	40.0 (48.0)	100

overtake the US as the biggest worldwide economy much sooner. Especially the growth of installed systems has a very positive effect: as will be shown in a later chapter the increased global cumulative volume – to which China makes a large contribution – will be a solid basis to further drive down the price for wind energy machines thereby making wind power economically favorable to conventional fossil and nuclear power generating systems.

4.3 Solar Thermal Collectors and Concentrators

4.3.1 Historical Development

The power of concentrated sunlight is already described in an ancient Greek tale, where the famous Archimedes is said to have defeated the Roman fleet in 213 BC by using glasses to concentrate solar rays and to set the Roman ships on fire. However, this did not help him and a year later he was killed by a Roman soldier before Syracuse was conquered by the Romans.

In 1767 the Swiss scientist Horace de Saussure developed the first cone shaped solar collector for cooking purposes. This was a glass covered "hot box" design which made use of the discovery that sunlight passes through glass, gets absorbed at surfaces painted black inside the box and the resulting heat cannot escape through the glass, thereby warming up the box to more than 100°C. This is more or less the principle of solar thermal collectors today.

In 1816 Robert Stirling developed a hot air engine, the principle of which is still elaborated today in the Stirling cycle engine. The air contained in a cylinder makes four cycles utilizing the energy provided by an external heat source: heating, expansion, cooling and compression to produce rotational motion. There may be a revival of this technology in the coming years together with solar dish systems which I will describe later.

Almost 100 years after the hot box experiments, Augustin Bernard Mouchot continued with this development and in 1872 he presented a solar powered steam engine using a truncated cone reflector (inverted lamp shade) with ½ horse power (hp). A few years later, in 1878 he used a 4m diameter mirror and 80l boiler to make ice using solar power, which was quite a sensation in those days in France during the Paris exhibition. Other developments during this time occurred in Egypt where John Ericsson developed a parabolic trough in 1883 for pumping water, which was very similar to today's Solar Thermal Concentrators. In 1878 in Bombay, William Adams built a central receiver which contained most of the features of modern prototypes in this technology.

Unfortunately, these efforts did not lead to commercialization although the underlying ideas of why more solar should be used were quite similar to today's considerations. In the Victorian 18th century some concerns were raised about the availability of coal to power the energy needs for the industrial revolution which were very similar to today's fears about peak oil and gas. The fundamental reason why solar thermal did not make it in the past but will be able to take over today is that in the early 20th century, oil became a much easier to handle secondary energy form especially to satisfy the needs that would arise in the upcoming First World War. Today, we can build on a portfolio of already proven renewable technologies with very competitive prices with the prospect of further price decrease compared to the increasing prices for conventional energy sources. A nice overview of this historic development is given by Magdi Ragheb [4-8].

The use of solar thermal today is split into two fundamentally different applications: firstly, domestic hot water and heat generation through flat plate collectors or vacuum tube collectors and secondly, solar thermal concentrating systems for electricity generation. The first category uses heated water at temperatures in the range 80 – 120°C for domestic use and 120 – 300°C for process heat for industry, while the second one heats special fluids to temperatures of 300°C to 600°C and a connected heat exchanger produces steam for electricity generation. Both categories can be combined with heat storage systems ranging from daily cycles up to seasonal storage of heat.

4.3.2   Solar Thermal Collectors

Two different designs for solar thermal collectors have evolved today: flat plate and tube collectors. The first category uses about $2m^2$ of specially coated metal sheets as an absorber. The coating should have a high absorbance for the solar spectrum and at the same time a low emissivity in the infra-red (heat radiation). The coating can be applied via vacuum deposition processes (e.g. sputtering) or via electro-chemical coating. Until recently the metal sheets were made of copper. To transfer the heat from the coated metal sheet, a metal coil is welded to the absorber where a fluid circulates which consists of water plus e.g. glycol to prevent freezing of the heat exchange medium at low temperatures – especially at times without sunshine. The absorber is normally placed in a box covered by a glass sheet.

The second category uses glass tubes (typically 2m long with a diameter of 50-100mm), where absorption again occurs via a coated metal stripe situated in the glass tube or through a second coated glass tube with a smaller diameter through which the heat exchange medium is circulated.

For better insulation, the volume between the outer and inner glass tube is evacuated ("vacuum solar thermal collectors"). Several glass tubes are mounted in a box and interconnected.

The system topography is also divided into two different categories: thermosyphon and external storage tank systems. The first system has the storage tank integrated and mounted just above the box containing the absorber. This simple approach utilizes the fact that hot water climbs up into the tank and then flows down through gravity when needed in the household. The disadvantage is the ugly look as there is no chance to integrate these tanks nicely and homogeneously into or onto roofs. If used on flat roofs which are common in many southern countries this concern is of less importance. The second system, where the tank is normally situated in the basement of houses must integrate an electrically operated pump which circulates the heat exchange medium.

Warm water for a household: Typical parameters for the supply of hot water during spring, summer and autumn in a typical single or multi-family house are as follows: collector area about 6m² and a 300l storage tank. Prices in 2012 were about €4,000 for flat plate and €7,000 for tube collectors. A typical cost split is 50% for the collectors, 25% for the storage tank and 25% for the pumping system and installation. It is important to note that the efficiency for a traditional oil or gas fired hot water supply is typically at its lowest in summer times where (almost) all warm water can be supplied by the solar system.

Solar thermal collectors for warm water and heating support: In cases where a household would also like to use the solar water heating as a heating support especially in spring and autumn but also partially in winter, the system must be bigger in size. Typically 15m² collector area together with an 800l storage tank should be installed. Prices are in the range of about €10,000 for a flat plate system. In order to make best use of the solar heated water for heating purposes it is desirable to have under-floor heating instead of radiators.

Technology: the major cost component for the flat plate collector is the absorber with ~50%. This used to be copper (Cu) in the past, but with the substantial Cu metal price increase (€3/kg to €7/kg from 2005 to 2011 at the London Metal Stock exchange, LME) the coated 0.2mm Cu sheet will be replaced by a coated 0.5mm Aluminium (Al) sheet, which should result in a 35% cost reduction for the absorber. The next step will be the replacement of the Cu piping with Al pipes. In the latter case, appropriate corrosion protection must be implemented. It is important to note that R&D is important for further cost decrease despite the general belief that flat plate collectors are low-tech products.

Thermal power instead of collector area: in public debates the measuring of solar thermal systems in m^2 collector area does not reflect the "real power" of such devices. Therefore an attempt was made by the IEA [4-9] to convert the collector area into the corresponding power at peak sunshine. The proportionality factor was determined at 0.7 kW_{th} for every m^2 of collector area. For the conversion of thermal power into the energy produced within a year as a mean average, the bank Sarasin assumed an energy output of 700 kWh for 1kW_{th} (more in the South, less in the North, proportional to the average global solar irradiation). A more general conversion function for various collector types and irradiation levels is given in [4-10].

The future of solar heating: While it is important to highlight the immense energy saving potential in today's times where almost half of the secondary energy of countries is used for heating purposes, one must remember that in future there will be an interesting competition between alternatives in serving a specific energy need cost effectively and energy efficiency measures. In the case of solar thermal needs, even today in most houses, we have the situation that due to insufficient wall and window insulation most of the annual secondary energy is used for heating. As long as this situation continues, it makes a lot of economic and environmental sense to invest in solar thermal systems. For new future houses (well insulated with less than a tenth of the heating requirements of the old generation, it may be useful to consider a different approach from a total cost of ownership point of view. For instance, the complete piping and central heating boiler assisted by the thermal collector system could be avoided by providing the small amount of residual heat through electrical heating – but only if this electricity originates from renewables, including the electricity for heat pumps. As today's investments in solar thermal systems will pay off in less than 10 years and since the replacement of existing houses with new ones is not likely to occur soon, one should encourage every home owner to invest in solar thermal systems for the time being. However, when looking at the longer time frame towards 2030 or even 2050, the dramatic decrease in required energy for heating and cooling due to proper insulation and the possibility of having water heated by electricity (resistance heating or via heat pump) should be taken into account and a simple extrapolation of today's thermal energy needs to the future may result in fundamental errors and the total of installed areas for thermal collectors would be much too high.

Regional installations: it is interesting to see how the use of solar thermal systems developed in different regions in the world in the past and where the biggest momentum can be observed. As summarized in Table 4.4 it is again China which has installed most systems in the past with almost

Table 4.4 Cumulative and annual solar thermal installations for the TOP 10 countries worldwide in 2010 [4-11].

Country	Cumulative solar thermal installations end 2010		Annual solar thermal installations in 2010	
	[million m²]	[GW$_{th}$]	[million m²]	[GW$_{th}$]
China	194	135.8	49.0	34
EU 27	21.6	15.1	3.1	2.2
Turkey	20.6	14.4	1.7	1.2
Japan	6.2	4.3	<0.1	<0.1
Brazil	4.4	3.1	0.5	0.4
Israel	4.0	2.8	0.3	0.2
India	3.1	2.2	0.6	0.4
US	2.7	1.9	0.2	0.1
Australia	2.5	1.7	0.4	0.3
Taiwan	1.9	1.3	<0.1	<0.1
RoW	~10	~7	~2	~1.4
Total	~271	~190	57.9	~40

200 million m² (corresponding to 136 GW$_{th}$). Almost ten times fewer installations were realized in the EU27 and Turkey, which both have a similar volume. The next group of 7 regions has an additional factor of between 3 and 10 times less, although there could be many more installations based on population and heat requirements: the US should have at least the same as the EU27 and India and Australia could also have a much higher installation base. Similarly, the annual installations in 2010 show China as the clear front runner with almost 50 million m² (34 GW$_{th}$), followed by the EU27 with 3.1 million and Turkey with 1.7 million m², with the other regions showing much less. When using an average of 700 kWh/kW$_{th}$ there was a remarkable heat production of 133 TWh in 2010 with an annual addition of 28 TWh.

Unlike in the wind sector, in the PV and solar thermal concentrator industry there are not just a few big, mostly market-listed production companies but a huge number of small and medium enterprises, most of them – unsurprisingly – in China, but also in Europe. It is interesting to remember that the glass tube absorber technology which dominates the Chinese market was originally a development by Dornier in Germany many years ago. As there was no interest in this technology in Europe in those days, they licensed their technology to China.

4.3.3   Solar Thermal Concentrating Systems for Electricity Production

The first Concentrated Solar Power (CSP) plant, very similar in design to today's systems was the "Solar Engine One". It was originally developed by the USA inventor Frank Shuman by using a series of hot boxes which were replaced by parabolic mirror concentrators as suggested by Charles Vernon Boys. The system was built in 1912 in Al Maedi near Cairo, Egypt, and started operation in 1913. It is shown in Figure 4.6. It consisted of five parabolic concentrating reflectors, each 62m in length and 4m in width, oriented in a north-south direction together with a mechanical tracker system which kept the mirrors facing towards the sun throughout the day. The steam engine was shipped from the USA and had an equivalent of 55 hp, enough to pump ~27,000 liters of water per minute from the river Nile onto desert land. It was the First World War in 1914 which abruptly put an end to this technological success story: all system engineers left Al Maedi and a British army contingent dismantled the system to use the raw materials for war efforts.

The tremendous development of the internal combustion machine in the first half of the 20th century, based on burning oil, was driven by the needs of all transportation purposes – cars, trucks and buses, ships and planes. The increased findings of oil fields in these times led people to believe that besides oil for transportation and coal for electricity production nothing else was needed for energy supply. The introduction of nuclear power after World War II ("Atoms for Peace") empowered the belief that

Figure 4.6 First parabolic trough power plant "Solar Engine One" in Egypt 1914 [technical archive of Deutsches Museum, München].

fossil and nuclear could power the world for the foreseeable future. It is no surprise that the renewable technologies, like Solar Thermal Electricity, were forgotten.

It was notably the first oil price shock in 1973 that people realized that oil – and also the other exhaustible primary energy sources – were finite and, additionally, may cause problems to our climate by warming the atmosphere due to CO_2. This initiated a huge R&D effort, especially in the USA where national laboratories, like SANDIA, were given large amounts of money to look for "alternatives" – which was the preferred name for "renewables" in these days. A major effort was undertaken in the field of Photovoltaics, wind power and also Solar Thermal Concentrators. Europe also developed different forms of Solar Thermal Concentrating Systems. The block diagram for Solar Thermal Concentrators is shown in Figure 4.7 and basically consists of the Concentrating Solar Thermal Field and a more conventional power block.

As an option, one could add a conventional gas or oil heated steam production and/or a thermal energy storage to prolong the time of electricity production after sunset.

The different technologies for the Concentrating Solar Thermal Field can be seen in Figure 4.8 which shows the state of the art parabolic trough as well as dish and solar tower systems.

Technology: The most matured solar thermal power technology is the parabolic trough system. The rough cost split for such systems is 45% for the solar field, 20% for the power block and Heat Transfer Fluid, 25% for the Balance of Plant and others, 10% for services and site work. Obviously

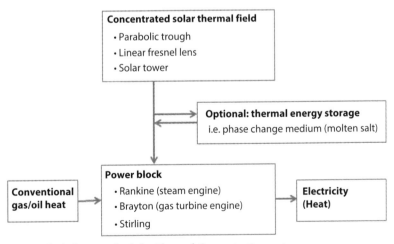

Figure 4.7 Block diagram for Solar Thermal Concentrating systems.

| Parabolic troughs | Dish | Power tower |

Figure 4.8 Various forms of high temperature electricity generating systems.

there is room for technological improvement for the solar field in both directions, which could reduce the cost of the running meter of the absorber and concentrating network as well as the cost per power unit by working at higher temperatures because of increasing efficiency in the power block. For the remaining 55% of the cost, there can unfortunately only be a cost decrease through increasing efficiency due to matured technology in the components used (turbine and associated components for producing electricity).

For the solar field there are two major components which can be optimized: the concentrating mirror and the receiver, which determines the efficiency of the system together with the heat transfer fluid.

Today the mirrors consist of 4-5 mm glass with sputtered silver to form the mirrors. These have a weight of ~ 10 kg/m^2. In the future, UV stabilized mirror film (ReflecTech^{TM}) laminated on aluminium may be used with a much lower weight of ~ 3.5 kg/m^2, thereby minimizing the structure cost. An additional cost decrease could come about through the use of small stripes of Fresnel lenses below the receiver which would further decrease the structure and tracking cost.

The heart of a STC system is the receiver. The state of the art today is produced by the company SCHOTT. With their PTR^{TM}70 receiver the following characteristics are obtained: solar selective coating on stainless steel (diameter 70mm) with high solar absorbance (>95%) and low thermal emittance (<10% at 400°C), surrounded by vacuum sealed glass (diameter 125mm) with glass transmittance > 96%, aperture length >96%. The bond between the glass and the steel tube at the interconnections between receiver tubes is important, as this may cause cracks due to temperature change from minus degrees Celsius at night to ~400°C at working conditions. SCHOTT was able to match the expansion coefficient of these two materials within the needed temperature range, thereby decreasing this defect. The great achievements of the receiver technology also present the problem of this technology: when looking at the technical numbers it can be concluded that the scope for a further increase in performance is only limited. There is no major technology development foreseen that would substantially decrease the specific system cost – mainly economy of scale and some material optimization can further decrease the cost per meter, including the interconnection.

Another limitation to the reduction of the specific system cost is the power block which is well developed – not only for this application. Hence the specific price ($/W) for this item will not decrease when increasing the volume of STC installations, but will develop in line with the world price for turbines in general.

After a quick decrease of the LCOE (levelized cost of electricity) in times when SEGS (Solar Electricity Generating System) I to IX were installed in the USA which added cumulative volume, further development has reached a point of saturation for reasons described above and lies today in the range of €ct18/kWh. This is about two to three times as high as for PV systems installed in similar locations. For the large green-field – or better yellow desert – PV systems, an important competitor to CSP, besides flat-plate PV, is the concentrated PV (CPV) system, which is well suited in large centralized systems in areas with high direct sunlight. Another advantage of PV systems is the absence of water needs which, however, is necessary for the cooling towers for CSP systems. Dry cooling towers can also be used with the disadvantage of substantially lower efficiency leading to a higher cost.

Another added value of STC systems is the storage of heat in phase change materials like molten salt to continue producing electricity after sunset. This enables a continuous production of electricity with stored heat from the sunshine hours during the night. The cost for storing heat per kWh electricity produced afterwards is substantially lower than the storage of electricity in electrochemical batteries. Depending on the price development for future batteries, this advantage may be an important added value for STC's with integrated heat storage.

A major decrease in specific cost could be envisaged if one could substantially increase the temperature of the Heat Transfer Fluid (HTF). Today's heat transfer oil limits this temperature to about 400°C thereby limiting the Carnot efficiency of the system (see also Table 4.8). If this could be replaced, either with water as a HTF for the production of steam directly, or with a still to be developed oil alternative, this temperature could be increased to 500°C or maybe 600°C thereby increasing the Carnot efficiency of the power block. This level of efficiency increase – if all other specific cost components could be kept at the same level – would then proportionally decrease the LCOE of STC produced kWh. This is the main reason why I personally still see a great future for this technology but only if this temperature increase of the HTF can be achieved. It will be interesting to see the performance of a recently started parabolic trough system ("Archimede" in Sicily) where water is no longer used to be heated in the receiver but instead a special molten salt is used. This enables an increase in the inlet temperature to above 500°C, thereby increasing the efficiency of the system. In addition, the same hot molten salt may also be used to store the heat in tanks for later electricity production at times when the sun is no longer shining. If these projects are not able to demonstrate this increase of efficiency it could well be that the parabolic trough technology may come to an end. This could then be the driving force to push the solar tower and dish technology as there will be higher temperatures anyway and this offers the potential to increase efficiency and in parallel to decrease the cost.

4.4   Bioenergy: Biomass and Fuel

An excellent overview of this subject can be found in WBGU [4-12]. Until recently (2005) bioenergy contributed ~60% to the 13% renewables share of primary energy. The breakdown of the ~14 PWh bioenergy is summarized in Table 4.5.

Although most discussions in our Western hemisphere are centered around the topic of biofuel, it is often forgotten that this is only a very small part (2.2%) and that the vast majority of bioenergy (86%) is used in form of wood for cooking and heating in developing countries. These ~12 PWh mainly from firewood could be replaced much more comfortably and

Table 4.5 Categories for bioenergy (~2005).

Category	% of total bioenergy	Energy [TWh]
Traditional (firewood for cooking & heating)	86.0	11,954
Modern bioheat	7.5	1,043
Biopower (corresponding to 45 GW)	4.3	598
Biofuel		
# bioethanol	1.8	250
# biodiesel	0.4	56
Total	100.0	13,900

environmentally friendly with new renewable sources like PV, wind and solar thermal. Biofuel only makes up a share of 16% of the rest of 1.9 PWh.

- For *biopower* there are many discussions on the topic of how to split the land use between agriculture for food production and plants for energy production (electricity, heat and transportation). There are many detailed studies which try to demonstrate that there is sufficient arable area for agriculture and also enough room for the production of biofuel and biomass for electricity production [4-13]. There are also many discussions derived from the problems with 1st generation biofuels and the introduction of a 2nd generation. This would use non-food lingo-cellulosic feed stock. While the 2nd generation of bioenergy does make sense this is not the case for the 1st generation as demonstrated below in a comparison between biomass production and state of the art renewable technology (PV and/or wind) on the same area of land: For electricity production from biomass we have numbers available [4-12] on how many MWh of electricity can be produced annually using the most advanced crops and technologies. These numbers are in the range of ~14, 17, 50 and up to 64 MWh per hectare for wheat, corn, sugar cane and best energy grass, respectively. Very often it is forgotten that the energy content per hectare (ha) is normally given as fuel energy. If we want to compare these numbers with electricity production from renewables, they must be divided by at least a factor of 2. In contrast we may compare this with the electricity production per hectare with today's PV technologies. If we conservatively assume a mere 10% of module efficiency, the modest solar irradiation in Germany with 1,000 kWh/kW$_{installed}$ and an area coverage of 50%, we obtain 1,000 kWh/kW$_{installed}$ × 1 kW$_{installed}$/10 m^2 × 10^4 m^2/ha × 0.5 = 500 MWh/ha which is more than 10 times the electricity produced from the best and fastest growing plants – and we did not even consider the water and fertilizer that would be needed for the biomass production. This factor 10 doubles if we calculate the production for sunny regions with twice as much sunshine as Germany, and doubles again to 40 if we install modules with 20% efficiency, which is already available today. Additionally, there is the possibility of a "double harvest" [4-14]: If one were to mount the modules high enough for a small tractor to pass below, one could even plant vegetables for local use below those modules (in regions like Germany, not in the desert, which would not have allowed crops to grow anyway).

- *For biofuel* a comparison between electric cars and combustion machines fueled by biofuel shows without any doubt that the (renewable) electricity pathway is by far superior. With the fastest growing plants we have 64 MWh/hectare or 64 kWh/(10 m²). Assuming 5 l/(100 km) (the energy content of 5 liter of petrol corresponds to 45 kWh) we can drive a traditional car for 140 km per year based on an area of 10 m². In contrast we have conservatively for a 10 m² roof mounted PV in Germany ~1,000 kWh (only 10% efficiency). With ~15 kWh/(100 km) for a 4 passenger electric car we can drive ~6,700 km with the same 10 m² – which is more than 100 times the distance with biofuel under optimal conditions! Using 20% modules the distance increases to 13,400 km which covers almost the annual needs for an average commuting car. In sunny regions with twice the output compared to Germany, the above numbers double again towards ~27,000 km/year which is more than 400 times the distance which can be covered with biofuel.

In summary, it can be stated that 1st generation bioenergy would be the wrong way to go for future energy supplies, both for electricity production and transportation.

4.5   Photovoltaics

This technology will be described in detail in the next two Chapters 5 and 6. In the context of the various renewable technologies it is, however, important to emphasize some of the specific aspects of this fascinating technology. While at the end of the 1990s it was widely believed that this way of producing electricity is the most elegant one, there was also a general understanding that this technology, which was at the time the most expensive one, would still need some decades and technological breakthroughs in order to become competitive. Due to forward-looking market support programs notably Feed-in tariff programs in Switzerland and Germany this technology could benefit from the well-known learning effect which is well recognized in other industries (Price Experience Curves for mass produced high-tech products, to be discussed in chapter 7). In essence, there is a close relationship between the price and the cumulated volume of such mass produced goods and for a given product there is a specific price decrease with every doubling of this cumulated volume. In the case of PV modules and inverters there is a 20% price decrease – also expressed as Price Experience Factor (PEF) 0.8 – with a doubling of cumulative volume. Due to the above mentioned support programs the market growth in the first decade of this century was an unforeseeable 50+% per year which boosted cumulative volume from 1.4GW in 2000 to an unanticipated 40 GW in 2010 and 100 GW by 2012! This implies a doubling of the cumulative volume by 2010 (6 times by 2012) which with the above mentioned PEF results in an expected price decrease of $(0.8)^5 = 0.33$, meaning that the price in 2010 is 1/3 of what it was in 2000 (~1/4 in 2012). While in 2010 the actual module prices were slightly above the expected number, we have seen an additional decrease in 2012. This is due to a very high production capacity increase, notably in China, resulting in severe overcapacity.

On the one hand this situation is very problematic for the manufacturers and there is a widespread belief that only a fraction of the biggest and strongest will survive. On the other hand, this gives system integrators the chance to compete against traditional electricity generating technologies in new applications and countries without support programs. An example is the addition of PV systems to an existing diesel gen(eration) set, thereby saving fuel resulting in overall cost reduction for the user ("fuel saver mode"). This will again boost the market growth and with new technologies for module production, as discussed in chapter 8, the production cost will allow for positive margins at today's prices.

In 1999 I sketched the PV generation cost as a function of time as shown in Figure 4.9. I restricted myself to the growth which was experienced in the decade 1990 to 2000, which was ~15% per annum. With this annual growth the cumulated volume would only have increased twofold, resulting in a price of only 2/3 in 2010 compared to 2000 (~40 €ct/kWh in 2010 compared to 60 €ct/kWh in 2000). The above stated 1/3 or 20 €ct/kWh was indeed already observed in 2010 instead of 2020, because a much higher growth rate of more than 50% per year was experienced.

This Figure was also often used when discussing the so called "grid parity". This is what occurs whenever the PV production cost in different locations (and for a given set of parameters, like WACC (weighted average cost of capital), etc.) becomes equal to the price the various customers have to pay to their electricity provider. As can be easily seen, there is no single point but rather four such intersection points (red dots) spanning an area of different times and locations when this grid parity occurs.

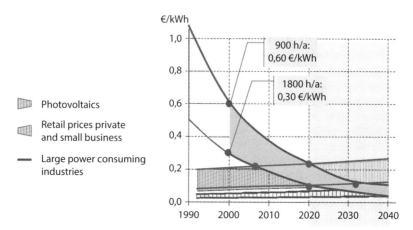

Figure 4.9 PV competitiveness (W. Hoffmann).

In terms of technology we will see in chapter 6 that there are many different technologies which will evolve for the different market applications, discussed in more detail in chapter 5. The important lesson for PV is that no known electricity generating technology has ever been able to demonstrate such a tremendous cost and price decrease in such a short period of time. This and the modularity of the components of a PV system from kW up to the GW in size will make PV one of the major pillars of the future electricity sector.

I have often heard, especially from economists, that market support programs are not necessary because the market forces will be the best regulator. These people have not understood first, the difference between strategic and consumer goods in a society, and second, the power of technological development associated with lowering the cost and price, if a well-tailored support program is able to provide the increasing cumulative volume of mass produced high-tech products. Both topics will be discussed in Chapter 5.

4.6   Other Renewable Technologies

This section deals with other renewable energy sources, important for the portfolio of a future 100% renewably powered world. We will not discuss these technologies in the same detail as solar, biomass and wind. This is based on the relative importance of the potential contribution within the various renewables as already described earlier in this chapter (compare with Figure 4.1). However, this should not be misunderstood as a dismissal of these technologies as not being important. They will play an important role in regional contributions which have boundary conditions that are very favorable to one of these technologies. But in the author's understanding, they will not be able to contribute to the 150 PWh challenge as much as the other ones, as will be discussed in more quantitative detail in chapter 9.

4.6.1   Hydropower

There are a good number of countries where conditions are favorable for this type of technology with a well suited landscape and good climate conditions. This enabled them to base a considerable part of their energy needs on the use of hydropower. The Top 10 countries together with the 10 biggest individual hydro power stations are shown in Table 4.6. As can be seen in the case of Norway, up to 98% of a country's electricity is supplied by hydropower. But also big countries such as Brazil and Canada satisfy a major part of their electricity needs through hydropower.

The global historical development of the annual electricity production over the last 46 years was a steady increase, starting with 0.9 TWh in 1965 and growing to 3.5 PWh in 2011 [4-15]. Interestingly, the increase over time was almost linear with approximately 0.5 PWh of additional electricity every 10 years. Let us take an average value of ~4,000 hours of full operational hours per year (calculated from the two stated numbers above: 3,498 TWh/850 GW ~ 4,000 h as a global average). Although this number seems quite low, it takes into account that in many regions the hydro stations cannot operate continuously, especially not in summer time. Combining the numbers indicates an annual increase of ~12 GW new hydro stations. In a publication by Horlacher [4-16] an estimation of the economic, technical and theoretical potential for the global hydro power electricity production was undertaken. The respective numbers are 8.2 PWh, 14.4 PWh and 39 TWh. If we take the same capacity addition for the future as in the past 4 decades, it would take 94 or roughly 100 years to reach the volume given for the economic potential for hydro-electricity. Another estimate was given in terms of price and electricity generation cost, which is summarized in Table 4.7 [4-17].

The large existing dams and attached power stations contribute the most to today's hydropower energy. This is normally called "old hydro" as most of these have already been in operation for quite a while. Although there are discussions on new very big hydro plants – for example in Brazil, Africa or Asia – there is growing concern related to the environmental and human damage this may cause. Additional projects like the "Three

Table 4.6 Top 10 countries and individual hydro stations as of 2011.

Country	Annual electricity production [TWh]	% of country's electricity consumption	Name of 10 biggest hydro power stations	Located in country	Installed power [GW]
China	694	22	Three Gorges Dam	China	23
Brazil	430	86	Itaipu	Brazil/Paraguay	14
Canada	377	61	Guri	Venezuela	10
United States	328	6	Tucurui	Brazil	8
Russia	165	18	Grand Coulee	US	7
India	132	16	Sayano Shushenskaya	Russia	6
Norway	122	98	Krasno–yarskaya	Russia	6
Japan	85	7	Robert– Bourassa	Canada	6
Venezuela	84	69	Churchill Falls	Canada	5
Sweden	67	44	Longtan Dam	China	5
Other	1,014	–	Other		760
Total	3,498	~17 (global)	Total		~850

Table 4.7 Investment and electricity generation cost for different sizes of hydro power stations.

System size	Investment per kW	Electricity generation cost
Up to MW	2.5 – 10 k€	5 – 19 €ct/kWh
Up to 100 MW	1.8 – 6.3 k€	4 – 11 €ct/kWh

Gorges dam" in China will become more difficult in the future. It should, however, be remembered that only ~10% of the currently installed hydro power base is accounted for by the very big hydro power plants, as can be seen in table 4.6. An important contribution to future energy can be made by the so-called "new hydro" which in most cases is a more decentralized power generation with smaller systems. In my understanding, the above stated economic potential of ~8 PWh is the upper limit for hydro energy.

Another important role which hydropower plays and will play in the future is the so-called "pumped hydro". Many countries like Switzerland are using cheap excess electricity (e.g. base load at night from neighboring countries) to pump water upwards and produce electricity when expensive peak power is needed. As peak power becomes less expensive with more and more PV (and wind) power produced, especially at peak times this business model no longer runs as effectively as in the past. However, for longer term storage this may become an important contribution when excess renewable Energy is to be utilized. Future business models will decide which portion of electricity will be stored using pumped hydro – which does not have high losses – and what that may be compared to the "power to gas" concept, which will be explained later. The latter has high losses originating from hydrolysis (electricity used to produce hydrogen) and converting it back into electricity if needed (by using fuel cells) which amount to roughly 50% losses (assuming about 70% efficiency for both hydrolysis and fuel cell).

4.6.2   Ocean Energy (Wave and Tidal)

Another way of using hydro power is the exploitation of the energy contained in the up- and downward movement of the waves, undersea currents and tidal hubs.

At places with high tide (e.g. Normandy in France with ~8m tidal range, with the spring and neap range as high as 13.5m) there have been attempts for many years to utilize the associated energy streams. One well-known example is the "Rance Tidal Power Station" in Brittany (France) with a 240 MW peak capacity which was completed in 1966. It has a barrage, 750m in length and 13m in height. In practice, the average power is about 96 MW, producing ~ 600 GWh of electricity per year. Interestingly, if we take the total cost in today's currency and divide it by the total amount of electricity produced, we obtain £3.3 billion / 27.6 TWh =~ 10 €ct/kWh [4-18]. Given the fact that this power station is situated at a very favorable location and all components are in a mature stage, this number shows that the difficulty with this technology is its relatively high generation cost. There is just one more tidal power plant of a similar size in operation today, which was opened in 2011 in South Korea. This plant is called Shihwaho, has a peak capacity of 256 MW and produces ~ 550GWh of electricity per year. Other running tidal plants are much smaller and located in Canada (20MW), China (4MW) and Russia (1.7 MW). There are dreams of a very large tidal power plant in the Bay of Fundy in Nova Scotia in Canada. If this plan could be realized in this place, which has the world's strongest and highest tides of ~15m, it could generate up to ~2,000 TWh – however, there is no estimate of the cost and the associated electricity generation cost for such a big project.

Instead of building dams, one could also utilize the underwater currents induced by the incoming and outgoing tide. First prototypes are being built and tested and it remains to be seen how durable these turbines will be in the harsh and salty environment, which will ultimately determine the economic viability of such systems. One interesting combination could be the coupling on the same pylon of a slowly moving rotor underwater and a normal operating windmill rotor. This could potentially help to reduce auxiliary costs like mounting structures and electricity network to transport the generated electricity to land.

Pilot projects are also being conducted with systems exploiting the wave energy. Swimming structures are mechanically connected and induce hydraulic pressure by moving up and down with the waves. The size of these pilot projects demonstrates the early stage of this technology, which is in the range of a few MW (like the Agucadoura Wave Farm in Portugal with 2.25 MW which was the first wave farm, started in 2008 and stopped by now).

4.6.3   Geothermal Energy

Geothermal energy utilizes the use of the internal heat of the earth originating from the magma below the earth crust. Magma, which is molten rock, comes to the surface when volcanoes erupt and streams of lava pour out. The basis for this power is the radioactive decay in the Earth's interior which is estimated to be in the range of 2×10^{13}W or 20 TW. Only about 40 to 80

Table 4.8 Carnot efficiencies "Eta" for various inlet temperatures at fixed low temperature at 25°C (T measured in Kelvin = 273 + x°C).

Inlet temperature [°C]	150	200	300	400	500	600
Eta $(T_h - T_l)/ T_h$ [%]	30	37	48	56	61	66

mW/m² reach the surface which is only a tiny fraction (<10⁻⁴) of the 1 kW/m² coming from the sun at normal incidence at peak solar radiation (or 100 to 250 W/m² continuous calculated average power). In some regions – Yellowstone Park, Iceland and many more places – hot water springs can be used to heat houses. The normal situation, however, is a constant temperature of between 10°C and 16°C in the shallow ground reaching the surface.

Once a geothermal power plant is up and running it continuously produces electricity which can be used for base load applications. Although the running cost is rather low, there are, however, important restrictions.

- high cost of drilling
  Depending on the geological formation, the cost of drilling the necessary holes is rather high and a major component for the generation cost for this technology. In a German ministry (BMU) study [4-19] the cost for a 5km borehole was estimated at €4 million.
- Low Carnot efficiency
  As the inlet temperature to the turbine is generally low in the range 200°C, the Carnot efficiency is also rather low. This leads to the fact that the specific investment cost for the conventional part for this type of electricity generation – steam generation, turbine, cooling tower – is rather high. For comparison, the Carnot efficiencies for a range of inlet temperatures are summarized in Table 4.8. Note that an inlet temperature of 150°C has only half of the efficiency of a state of the art turbine fired by fossil or nuclear produced steam.

Combining the different input parameters leads (with an investment cost of ~(2.5 – 5) k€/kW and 8,000 full load hours per year) to an electricity generating cost of (7 – 15) €ct/kWh.

4.6.4 Heat Pump

The utilization of surrounding heat (air or soil) with heat pumps requires a considerable portion of external energy. According to a study by the

German ministry BMU [4-20], heat pumps can be called hybrids between saving conventional energy usage and renewable Energy. To provide the same 100 kWh heat for a household, a comparison was made between a standard caloric value burner, an electricity driven heat pump, and a gas motor driven heat pump. While for the classical gas burner 112 kWh's are needed, the corresponding numbers for the electricity (using primary energy numbers) and the gas motor driven heat pump are 78kWh and 66kWh, respectively. It is interesting to see that at the time the study was conducted in 2004 the examples given for various external energy inputs did not recognize that if electricity was provided by renewables then this device was no longer a hybrid but a 100% renewable device. If we only take into account the secondary energy, then the electricity driven heat pump needs only 26 kWh, making renewable electricity the most efficient way of powering a heat pump. There is also the misleading guideline as to which "coefficient of performance" (COP) for an electrical heat pump should be used. This COP is defined as "useful heat/work input", which for our above example is 100/26 ~ 4 using secondary energy numbers for electricity, but only 100/78 ~ 1.3 when using primary energy data. For conventional electricity, the "work input" in form of primary energy is 3 times higher due to losses in the power station, therefore it is normally recommended that this COP for electrical heat pumps should be higher than 3, while for gas engine driven heat pumps it only has to be above 1.1. With renewable electricity, the same lower limitation should be applied.

5

PV Market Development

5.1 Strategic and Consumer Goods in Society and Why Strategic Ones Need Initial Support

Before we start explaining and understanding the growth of the PV or other renewables sectors, we first have to ask why market support programs should be introduced for technologies which have a much higher cost at their start – like electricity from PV or other renewables – when the same product can be produced much more cheaply with conventional technologies – like gas, coal and nuclear powered plants. The product kWh is no different depending on which power generating system was used. So, why then should politicians facilitate support programs that society has to pay for?

5.1.1 Consumer and Strategic Goods – a Message to Economists

We must differentiate between two distinct products within a society. I like to call them "consumer goods" and "strategic goods". Figure 5.1 gives some

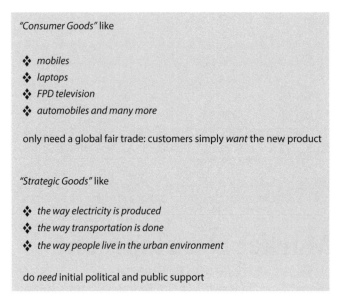

"Consumer Goods" like

❖ *mobiles*
❖ *laptops*
❖ *FPD television*
❖ *automobiles and many more*

only need a global fair trade: customers simply *want* the new product

"Strategic Goods" like

❖ *the way electricity is produced*
❖ *the way transportation is done*
❖ *the way people live in the urban environment*

do *need* initial political and public support

Figure 5.1 Consumer and strategic goods.

examples for each category. It is obvious to me that the consumer goods should not be used to influence the respective production industry with industry-political actions – although I know that in many cases this is unfortunately happening. Since people simply want the new car, the new mobile phone, the new house and many other consumer goods, we should have a level playing field for the global industry to create efficient competition for the benefit of the customers. Free trade, no state interference and a worldwide contest is the right way to go, and here I am in agreement with some of the liberal political principles. In contrast to this, there is the way strategic goods should be handled. Just some examples:

- **… the way electricity is produced**
  The product sold to the customer is a kWh electricity. The product kWh will be not different depending on the technology producing electricity, but the cost will be different. As a consequence, each utility company would just adopt the approach with the lowest cost in order to have an advantage over their competitors. A new technology would have no chance to be utilized as long as it was more expensive than the existing methods. And any new technology entering the market will start with high cost and high price at low volume production. Without the hundreds of billions of dollars spent on support and subsidies, the nuclear industry would never have had a chance to enter into utility scale competing with coal and oil 40 years ago. But at that time the whole world believed

that this new technology could really be the future and would provide endless and cheap electricity and therefore all the relevant states spent the necessary billions so as not to lose out against the others. By now we should know better!

- **…the way transportation is done**

  I remember when in the mid-1980s politicians in Europe pushed for a legislation on the catalyst to remove HC, CO, NO_x and – for diesel cars – unburned carbon. An additional positive side effect was the removal of lead in the petrol as this would have been a poison to the catalyst. The first reaction by the European carmakers was an outcry, although they should have known better, since the introduction of the catalyst had been successfully carried out 10 years beforehand in US and a few years beforehand in Japan. Without the interference of politics, we would most probably still today have the dirty exhaust by the millions of cars on European streets; at the very least introducing the catalyst would have been delayed by many years which would have caused a lot of damage to the environment and to the health of people.

- **… the way people are living in the urban environment**

  Without additional incentives most existing houses would take much longer to get refurbished in the way which was described in the preceding chapter: insulation, low-e windows, heat pumps, solar thermal, PV systems and much more. The new building directive in Europe which will enter into force in 2018 pushes even harder to reduce the traditional waste of energy.

To summarize: if for strategic goods within a society drawbacks are encountered for the traditional way in which a strategic good was provided so far, it is necessary to provide either a legal framework (like with the catalyst) or a support for the new industry (like nuclear 40 years ago, or renewables today) either by subsidizing the industry or – as I will show – through a much better way: supporting the customers in buying the new technology. It is part of the true leadership of politicians to recognize the development of new and better ways to serve important strategic goods in a society and to provide the appropriate support to bring this new technology into the mainstream.

5.1.2   Market Pull Versus Technology Push – What is the Best Support Program?

In the preceding chapter it was already highlighted that technology push programs like for the wind energy in the US in the 1970s or Germany with its GROWIAN in the 1980s have not been successful in introducing a new technology. Millions of dollars in R&D money were given to big research organizations which worked together with research divisions from – most often – big companies and who made the mistake of using a scaling factor which was much too high compared to the state of the art technology. Such research projects in most cases resulted in prototype machines which did not make their way to volume production (for examples see the wind energy chapter). In contrast to just funding big research projects at institutes and companies, one should additionally support the investor of the device, for example in PV or wind, in producing the product under consideration e.g. electricity. There are two possibilities: either the state can provide support for the investment in form of an investment subsidy and/or tax credits, or by paying a fixed tariff for that product over a defined period of time. The first scheme was widely used in the US and Japan (until recently); the second was first introduced in Germany as early as in 1991 (feed-in law for renewables in general plus an investment subsidy of 70% for PV systems) and was adopted for the different renewable technologies in 2001 with the EEG (renewable Energy Sources Act) without investment subsidy.

There are several drawbacks with the investment subsidy scheme as experienced in the past. First, a maximum price value must be defined for this investment. In the case of PV in 1991 this maximum price was determined by the funding ministry at 13,500 € per kW installed. It is no surprise that it did not take too long before all over Germany the specific price for a PV system was at or around that number – for lower as well as for better quality systems! This does not really create a competitive market. Another finding was the investor's loss of interest in long term performance, especially private investors. Once the subsidy had been received, many did not pay attention to the best continuous functioning of the complete system, including regular maintenance.

These disadvantages can be overcome when the support is given in the form of a properly calculated fixed tariff over a defined period of time. In the case of PV the principles were defined along the following philosophy: the period of time was determined at 20 years, also determining the depreciation time of a PV system. This should guarantee good quality of the components as well as of the workmanship. The proper calculation was done by taking the above depreciation time, the financing conditions at the time for the investment, and an allowance to make a decent profit on the invested capital over the 20 years. "Decent" should mean that the return on investment should be in the range of more than ~6% but less than ~10% per year. The numbers were not fixed but were seen as a gentlemen's agreement in the public/political discussion: more than 6%, because otherwise investors would take other investment decisions and less than 10% as the money for this tariff is collected from all electricity payers. As the tariff is paid for 20 years, the investor has a great interest in maintaining a properly working system for the entire period, and it also makes investors look regularly at whether the electricity production is really what it should be. There are web pages today which show for each postal code how many kWhs per kW should have been produced within the actual time period for a standard system. Quality is also of major importance because the whole system must at least work for 20 years with only little degradation. The last, but definitely not the least important argument for this type of support is that it creates a highly competitive environment. Because the investor in the PV system gets the support, he looks at the market place where he can buy the product at the desired quality for the best price. It is therefore no surprise that in Germany we have the worldwide lowest prices for PV systems per kW for the different applications.

As will be described later, the *FiT (Feed-in tariff) program* which was introduced in Germany in 2001 was able to initiate a growth of the PV market which was very successful, both in terms of quantity and also in terms of price decrease. An important aspect is also where the budget to pay the FiT should come from. If it is taken from a ministry's budget (coming from the tax payer) it is very dangerous, as tight budgets in countries will always initiate a new debate every year on how much can be afforded in the budget for the next year. If, however, this budget is distributed among all electricity users, this represents a much easier way forward – as long as the increase in the kWh-price is reasonable and accepted by the general public.

A short remark on another support scheme called *quota together with certificates*. Here politics defines a fixed share of renewables and utility companies are obligated to fulfill this with the evidence of certificates. These certificates are tradable. Utility companies are either able to produce the needed share themselves or they buy certificates from operators of renewable energy systems. If the company is not able to provide enough certificates relative to its produced electricity it will be fined. Advocates for quota systems often argue that this support is more market conform compared to FiT's because there is some trading involved.

FiT's basically have no limit for the share of renewables – which is a great advantage over the politically fixed volume for quotas. The incentive for the investor is the guaranteed price for a fixed time (normally 20 +/- 5 years) dependent on the technology and region. Furthermore a digression of the FiT as a function of time is foreseen to account for technological progress.

For countries which have introduced either a quota system or FiT-program for the same technology, in this case wind energy, the number of installations and also the resulting cost figures are compared in Table 5.1. If countries with similar electricity markets are compared (e.g. Germany with Great Britain, Spain with Italy) there is a remarkable increase in size by a factor of about 4 to be seen for countries with the FiT support scheme. Also, the resulting price level is considerably better for FiT-countries by more than 50% on average – and this for countries with comparable wind conditions. This can indeed be taken as clear evidence of the superiority of FiT-programs over quota systems.

It is interesting to follow the political discussion in many EU member states up to the highest levels in the EU Commission where supporters of quota systems would like to install a so-called "harmonized support scheme for all renewables within the EU27". What may sound quite nice turns out to be the best strategy to prohibit the quick growth of renewables. The ultimate goal for them is to install only those renewables which at a given place and time are the cheapest ones, namely centralized and large systems. Especially the small and decentralized systems like PV on roofs would become obsolete, because no utility company would make any investment and private households would no longer benefit from such installations. As one Commissioner always argued: wind systems should only be installed in places with high wind speed – preferably off-shore wind parks – and solar systems only in those regions with the highest insolation – preferably in Southern Europe. In both cases it may take many decades until a significant share of renewables is established, because off-shore needs some

Table 5.1 Comparison of wind power price and cumulative volume in countries with FiT's and quota at the end of 2011 (source: EWEA, 2012 and EREF, 2009).

Country with FiT	Cumulative wind power capacity [GW]	Price [€ct/ kWh]	Country with quota	Cumulative wind power capacity [GW]	Price [€ct/ kWh]
Germany	29.1	8.9	Italy	6.7	14.9
Spain	21.7	7.8	Great Britain	6.5	10.8
France	6.8	8.2	Poland	1.6	11.4
Portugal	4.1	7.4	Belgium	1.1	14.2
Denmark	3.9	7.1			
Ireland	1.6	6.8			
		Ø 7.7			Ø 12.8

additional time to solve the technological challenges and Solar in Southern Europe would need Super Grids to transport the generated electricity to the consumers in the rest of Europe through a number of countries.

My conclusion from these facts is that those pushing for a quota system rather than a FiT system are actually fighting against the cost effective large volume introduction of a competitive portfolio of renewables over traditional energy technologies.

Today, there is a highly controversial discussion around the globe on the following topic: For the strategic product "renewable electricity kWh", a country has decided on a support scheme for PV with a feed-in tariff as just described. The question arises whether it is appropriate to give preference to locally produced components (also called "local content" for modules and inverters) as long as the country's population is paying for the growth of the market and the politicians have to take the responsibility.

One extreme is found in the state of Ontario in Canada, where a law was put in place that required 60% of local content for the system components from 2011 onwards. This implied that most of the value added work for a module or an inverter had to be done locally, which had resulted in a number of new production companies for these components. The disadvantage of this approach was twofold: as production cost for the relatively small volume factories was significantly higher than the state of the art production facilities, the price at which the components were sold to the customers also increased, which resulted in a much higher installed system price, requiring a proportionally increased feed-in tariff; in 2012 the FiT in Ontario was more than three times as high as the one in Germany although the average global irradiation is more than 25% higher, which means that FiT in Ontario should be 25% lower (annual irradiation (900 – 1,200) kWh/m² in Germany compared with (1,300 – 1,400) kWh/m² in Ontario. However, the feed-in tariff in 2012 was (13.5 – 19.5) €ct/kWh in Germany and (64.2 – 80.2) CADct/kWh which corresponds to (49 – 62) €ct/kWh). Another aspect is the violation of the WTO rules which are accepted by many countries in the world. It remains to be seen whether the legal actions already taken by other countries (e.g. Japan) will put an end to this type of local content.

In Italy, another interesting scheme for local content has been proposed, which also aims to push the local production industry. The approach taken was a bonus to the FiT that was paid if the product was produced in Europe, where the bonus is 10% higher if more than 60% value adding to a component is done in Europe. Also, France introduced a bonus scheme late in 2011, which adds 5% to the tariff if the module is fabricated in Europe and an additional 5% if also the solar cell is fabricated in Europe. This scheme does not prohibit the import of any components and the level of the bonus is quite reasonable – it will be interesting to see how this scheme develops, although it in principle also violates the WTO rules in general but is much "softer" than high import duties or the Ontario model.

In all WTO countries companies within an industrial sector have the right to ask the responsible legal entity (i.e. DG Trade in Europe, the Department of Commerce in the US) to determine import duties for products coming from a particular country if an infringement of anti-dumping or anti-subsidy rules can be proven. This was done in the US in 2012 for PV modules containing solar cells produced in China. There is an ongoing procedure in Europe (to be concluded at the beginning of December 2013) for wafers, cells and modules produced in China.

5.2   PV Applications and History

There are many different customer needs which can be effectively powered by PV. The optimal PV product is by no means a "one size fits all" solution. However, the different PV technologies, which will be explained later in more detail, can help to make nicely custom tailored products. Another important aspect for the many applications is the different parameters which have to be met when a specific product is optimally developed. This can be best understood by taking a closer look to Figure 5.2. The four rows show for two typical examples each of PV systems for on-grid (that is PV systems integrated into the electric network), off-grid (stand-alone systems, not attached to the grid), consumer (ranging from powering watches and calculators through garden and house appliances to integrating them into other products like sun roofs in cars) and high efficiency products for special applications. The timely development when the major applications entered the market is shown in Figure 5.3.

Below each picture the specific focus for the different applications is given. Starting at the top right hand we can see the first commercial use, namely solar cells to power satellites. This was first done for the Vanguard I satellite, launched in 1958 in the US. Prior to that, the satellites had been powered by atomic batteries, which had a clear disadvantage to the lightweight solar cells due to their high weight (Promethium 147). It was and still is the *weight per power (g/W)* customer focus which drives technology for PV in satellites. First, greater and greater efficiency was developed with crystalline solar cells which were increasingly switched to the substantially higher efficient III-V solar cells (GaAs based) in the 1990s, which today surpass an efficiency of 40%. An interesting spin-off is the introduction

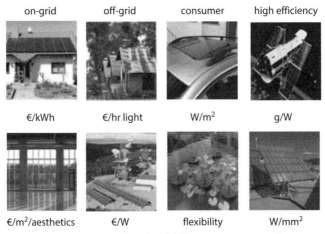

on-grid	off-grid	consumer	high efficiency
€/kWh	€/hr light	W/m²	g/W
€/m²/aesthetics	€/W	flexibility	W/mm²

Figure 5.2 PV systems serving a multitude of different customer needs.

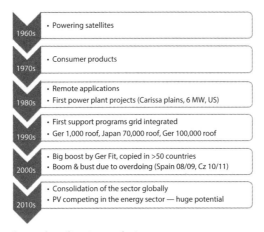

Figure 5.3 PV market and application evolution.

of concentrated PV (CPV) in recent years, where these highly efficient solar cells are mounted on a two axis tracking device and each small cell (~5×5 mm²) receives solar light from a concentrating Fresnel lens with a size of ~100 × 100 mm², thereby reaching a concentration ratio of about 400. Obviously the customer focus in this case is the *highest efficiency per area (W/mm²)*. Concentration ratios can go as high as 1,000. A typical system is shown in the bottom right hand picture.

The row to the left shows consumer goods. This application was started in the 1970s and was first used in millions of calculators and other small appliances. Preferred technology here was an amorphous silicon solar cell of a few cm² area. The bottom picture shows a typical garden product swimming in a pond, where the round solar module drives a little pump which – proportional to the sun shine - gives a small water fountain.

The *aesthetically pleasing integration of the solar module into the body housing for the various (OEM, Original Equipment Manufacturers) products* makes thin film modules the preferred choice (round and different cell shapes for easy series integration, flexible modules to accommodate for bent body housings). The upper picture shows an application of a solar module integrated in the sun-roof of a car. The customer service here is a cooler interior of the car when sitting in the sun whilst parking. The module drives a ventilator, sucking cool air from beneath the car and thereby creating a considerable temperature drop inside the car which is at its biggest when it is most needed, because the more the sun is shining the quicker the ventilator turns. This is also the reason why for this application, where only a given and limited area is available for the module, the *efficiency of the solar cells (W/m²)* is a decisive factor.

The second row from the left illustrates two typical examples for off-grid applications. The bottom picture shows an industrial application, here the replacement of the diesel generator to power a remote telecommunication system. In former times there was the need for a regular visit by helicopter to fuel and maintain this system. With PV modules, batteries and a small diesel engine for the few times of year when the batteries are empty and there is no sunshine to charge them, much fewer visits are necessary, making the power supply much lower in cost per year. Other important industrial applications are the cathode protection of pipelines to prevent corrosion, off-shore navigation warning lights, horns in lighthouses, railroad crossings and many more remote systems where utility grid connection would be more costly compared to the PV system. Customer focus here is on *high quality and a long life at a competitive price.*

The other major application, seen in the top picture as solar home systems (in Indonesia), is to bring electricity to the billions of people living in rural areas in emerging economies and developing countries. Without PV they would not have access to light and communication tools for a long time to come. But with a solar home system as can be seen in the upper picture, the small module connected to a small battery can provide this service at a lower cost compared to burning candles to have at least a bit of light in the evening and instead of using dry batteries for listening to the radio. Adding the newly developed LED products which need much less electricity than former light bulbs there are so called "pico PV systems" available which combine a rather small PV module with such energy efficient lamps to cost effectively bring light to these regions and the people who live there. Also PV powered water pumping, both for drinking and irrigation, is less costly today as well as environmentally safer compared to diesel powered systems. The customer focus for this remote use is the *cost for the hours of light, which can be offered to the household, the liters of water pumped with such a system, the time one can be connected to the world via radio, television and internet.* In the 1980s this application was perceived as the major market segment for the coming decades. By connecting several houses together one can create a mini grid to efficiently power many remote villages. With today's prices of the components of such a system (modules, inverters and batteries together with energy efficient appliances to decrease the electrical power input) the old dream of bringing power to billions of people which we had back in those days, could become true in the coming years. Access to money is also important – even if the cost is only some 10s of $ it constitutes a major investment for a household in developing countries. This can be supported through proper micro-financing schemes despite the fact that some of these schemes did not always achieve what they should have – which is not the fault of the micro-financing idea, developed by Muhammad Yunus with his Grameen bank, but rather due to the greed of individual banks who mis-used the scheme by asking for interest rates which are way too high and thereby driving customers into ruin.

The first row shows typical examples of on-grid applications: the top picture displays a roof-top system, where the modules are mounted on top of an existing roof of a private house, and the bottom picture shows a so called Building Integrated PhotoVoltaic (BIPV) system. In order to have grid connected systems as a mainstream application, the customer is focused on €ct/kWh *generation cost.* This firstly allows a comparison with the household price which he has to pay for today's portfolio of power generating systems and at a later stage – actually not too far in the future – it will allow to compete with gas, coal and nuclear power generation. An important aspect for an additional customer need to be served in BIPV applications is the *price/m² and the homogeneous appearance* of PV products when integrated into a façade or overhead shading.

The market split for the market categories on-grid, off-grid and consumer from 1995 to 2010 is shown in Table 5.2 [Navigant, 2012]. One observation is that in 1995 there was a clear dominance of off-grid applications with more than 80%. This changed quickly because of the rapid growth of on-grid systems. Even the off-grid, consumer and high efficiency applications grew in this time interval at a rate of about 15% per year. Most industries would be happy if they had such a growth rate over a period of ten years. But the on-grid segment, unanticipated by almost anyone, grew at an average of 66% annually during these years and increased its market penetration from only 13% in 1995 to a dominating 98% in 2010.

Table 5.2 Historical development for market segmentation.

Year	Total annual market (MW)	Off-grid (%)	Grid connected (%)	Consumer (%)
1995	72	82	13	5
2000	252	47	51	2
2005	1,408	17	83	1
2010	20,000	< 2	98	< 1

5.3   Historical PV Market Development

Early in the 1980s, the US was the engine for PV production and deployment. As early as 1983 a 6.4 MW installation by the market leader at that time Arco Solar (later Siemens Solar, Shell Solar and then integrated into Solar World (crystalline Silicon business) and Saint Gobain (Thin-Film business)) was built for the utility company Pacific Gas and Electric (PG&E) in Carissa Plains, California. Another subsidiary of the oil company Amoco, Solarex, was also one of the biggest producers in the early days (their slogan in these times was 'we are not the biggest, but the best'). When BP merged with Amoco the two subsidiaries Solarex and BP Solar were combined into BP Solar, which was for quite a while one of the Top 3 module producers worldwide (see also Table 5.3). The latest news is that BP Solar is just about to withdraw from the manufacturing of solar modules. With the exception of Total, who came from the downstream installation business and recently acquired Sun Power, one may ask why most of the big oil giants – Exxon, Shell, BP, ARCO, Mobil – who invested in this PV technology in the early days are now divesting. A couple of thoughts are:

- The big rush of oil companies into PV started after the oil crisis in the 1970s. Many of these companies were looking for new business opportunities related to energy
- The investment in the early days was comparably small for such big and rich companies and management was used to spending a couple of hundred millions of dollars on boreholes which were empty but at the same time finding ones which brought in a lot of money – PV was possibly seen as an either/or enterprise
- For the solar subsidiaries, the management positions in most cases were staffed in a similar way as in the mainstream business: job rotation of CEOs every few years (during my time as CEO of ASE and later of RWE Solar I have seen a handful of CEO colleagues in each of these solar companies)

- The mass production of PV modules and its subcomponents requires a different background in the executive mother company compared to what the mainstream business is doing (similar arguments apply to utility companies)

As a consequence it is not surprising that either dedicated newcomers or heavy weight mass producers are the ones to dominate the production arena in the future.

But now back to the further growth of the PV-market. After the first market push it was only California who made further installations, but on a small scale. Also in Italy a number of larger projects were installed in these years of the 1980s by the utility company ENEL, for example the 80 kW system on Vulcano Island or the remarkable 3.3 MW Serre plant. Similarly, the German utility RWE added a 300kW demonstration plant in Kobern-Gondorf in 1988, primarily aimed at testing the state of the art for module and inverter products but also looking at how to integrate PV into a nice-looking green field surrounding (they hired a botanist to help rare species of animals and plants to settle within the solar field). Another 1 MW test field was completed in 1994 in Toledo (Spain).

Projects such as those mentioned but also in general are no basis for a stable or even growing market. A first demonstration of integrating PV systems into private houses was the 1,000 roof program in Germany which started in September 1990. Because of German reunification this number was increased to a total of about 2,500 PV systems, each with a size of between 1 and 5 kW. The average size for all systems installed until 1993 was about 2.5 kW, so the total installed capacity for this demonstration project was about 6 MW. The support was given to the investor in form of an investment subsidy of 70% (up to a maximum of 13,500 €/kW installed price). An additional important cornerstone was established: in order to have a smooth grid integration, a feed-in law for renewables was introduced to support the above described PV project and also to support the installation of several 100 MW of wind energy. Besides many important positive achievements this project had one major deficit for PV: the annual size was very small (~2 MW per year) compared to the annual global market of about 60 to 80 MW in these years. As a consequence, the industry – the PV manufacturers in Germany and other European countries – did not need to invest in additional capacities. Obviously an industry-political goal to push the manufacturing industry at home was missing.

This was very different in Japan. I remember in the early 1990s during the German "1,000 roof program" when officials from the Japanese Ministry for Economy, Trade and Industry (METI) visited Germany and to look in detail at how the support was organized in the German demonstration program. In 1993 the "New Sunshine Project" was launched and an incentive program was started in 1994 with an investment subsidy similar to the one in Germany. This "Residential PV system Dissemination Program" – better known as "70.000 roof program" – was oriented towards commercialization of PV and especially towards building additional production capacities in Japan. This was possible because the program size allowed for it. With an average of 3 kW it was a total of more than 200 MW which had to be installed. As a consequence, until 2002 Japan was stimulating and leading the global market as can be seen in figure 5.4.

The simple goal was that the big Japanese module companies should invest quickly to reach mass production earlier than other companies across the globe. This should result in manufacturing in Japan at lower prices so they might be enabled to efficiently win market shares in other solar markets in the EU and the US by forcing out of the market those competitors, who had not reached mass production levels, and were thus likely producing at higher cost. In the first years, while Japan was the biggest global market place, the Japanese companies would focus on building sufficient capacity to supply the home market. It is no surprise to discover that, in 2000, the Japanese manufacturers enjoyed a ~40% market share compared with ~30% for US and ~20% for EU competitors. The details for the Top 10 manufacturers are shown in Table 5.3. The situation was summarized in an article by Oliver Ristau [5-1] in 2001: "In the coming two to three years, a Japanese-style export offensive can be expected. Until then, as defined in the METI Research Program "New Sunshine Project",

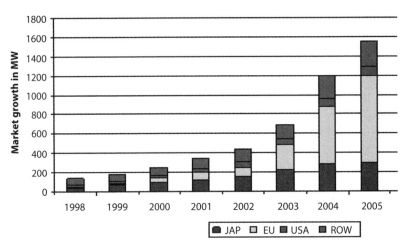

Figure 5.4 Regional market evolution until 2005.

Table 5.3 Top 10 module production companies in 2000.

Company	% share	MW shipped	Company	% share	MW shipped
Kyocera (J)	15	42	Photowatt (F)	5	13
BP Solar (US)	15	42	Sanyo (J)	4	11
Sharp (J)	15	42	Isofoton (Spain)	3	8
Siemens Solar (Ger/US)	11	31	Mitsubishi (J)	2	6
Astro Power (US)	6	17	other	19	55
ASE (Ger/US)	5	13	total	100	280

companies should have successfully taken the steps to mass production. Then what the Japanese had accomplished in other industry sectors could become reality: the flooding of foreign markets with low-priced, mass-produced articles." However, as careful as the considerations of the METI were, the reality in later years developed quite differently. One drawback was the phasing - out of the investment subsidy program for PV installations at a point in time which was too early compared to the generation cost which had by then been achieved. This resulted in a slowing of growth until 2005 as seen in Figure 5.4, including even a decrease of the annual installations after 2006 until 2008. Although the annual market installations could be positively influenced after the introduction of support measures (compare Figure 5.5), the dominance enjoyed by Japanese producers suffered in a first wave when German manufacturers surpassed them in size and even more in a second wave which witnessed the rise of a huge Chinese production industry strategically supported by the Chinese government.

After the end of the "1,000 roof program" in Germany there was only a small market increase, driven mainly by some municipality projects with some remarkable new ideas, like cost effective feed-in tariffs of about 1€/ kWh paid to investors in several communities and towns, like in Aachen, Hammelburg, Munich and several others.

It was at the start of the newly elected Red-Green Coalition in Germany in 1998 when it was realized that it might not be possible to quickly introduce a feed-in tariff but that another market pull for PV could be established earlier . In 1999 the so-called 100,000 roof program was started with the goal of installing 300 MW PV systems. At the beginning, investors were given a zero interest rate loan for investments in solar PV systems. This resulted in a monetary benefit at the time of not more than 10% when compared to a credit, customary in banking. Due to relatively small-scale financial support

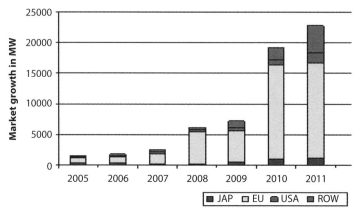

Figure 5.5 Regional market development until 2011.

the program was initially taken up rather slowly but accelerated quickly after being combined in 2000 with the Renewable Energy Sources Act (RESA = EEG). Until 2005, more than 90% of the European market, as shown in Figure 5.5, was in fact the German market. With the increasing German market, the year 2003 marked an important shift from Japan being the biggest market for PV installations until 2002 and Germany taking over this role in 2003 due to a much more generous support program for the investors in PV systems. At this point it is helpful to take a closer look at both the development of the feed-in tariff program as it had evolved in Germany since 1991 and how it originated. My own recollection of the chronology of this new and mind-set changing support program was as follows:

Among those who helped advance the renewable case, Wolf von Fabeck stands out. His contribution was important in regard to the question of cost-covering remuneration for electricity produced from all the different renewable technologies. After founding the Solarenergie-Förderverein Deutschland e.V. (SFV) headquartered in Aachen in 1986 he pioneered advocacy of this new form of market support in the late 1980s, with the objective of reaching 100% renewables as quickly as possible. Obviously he was also a forerunner in propagating a 100% renewably powered world which now, 20 years later, appears to be a topic taken seriously by more and more people (including the author).

In order to demonstrate the maturity of renewable energy theory and practice, two demonstration programs -"1,000 roof PV" and "100/250 MW wind" - were introduced in Germany in 1990. As already described, most of the financial support for the PV roof program was a 70% investment subsidy (with all the downsides mentioned). I remember the times when we had discussions about what else was needed to make this demonstration program a success. A bi-annual meeting was established ("Glottertal meetings") in the late 1980s where industry, institutes and ministries regularly came together. In the late 1980s a new director in the department of the German ministry responsible for renewables was appointed (Dr. Sandtner). He did not come from the technology sector (as most of his predecessors) but from a legal and legislative background. The technology-oriented community at the time had some reservation over his appointment as we – wrongly – thought that we needed more technical expertise to further develop our industry.

However, fortunately for our industry, he introduced new ideas. An important feature was that a law was passed by government stating that renewable kWh had to be preferentially accepted into the grid, plus a feed-in tariff of 8.3 €ct/kWh for all renewable technologies. While this feed-in tariff was only a tiny fraction of the PV generation cost of about 110 €ct/kWh in those days, it was a stimulus for growth in the wind industry in Germany which already had a comparable generation cost on a good wind site. Hence, more by chance than by design, for wind energy, this "Stromeinspeisegesetz – STREG" (electricity feed-in law) in 1990 was already a cost-covering remuneration for electricity from renewables on windy sites.

It was in the small town of Burgdorf in Switzerland that a law was passed in 1991 to remunerate PV electricity at a rate of 1CHF/kWh for 12 years. Fortunately, this legislative framework was positively supported by the director of the local municipality (IBB, Industrielle Betriebe Burgdorf). As a consequence, PV power installed per capita – suggested by Thomas Nordmann as an easy to remember illustrative number – was much higher in this town compared to the rest of Switzerland. Other local municipalities in Germany, like Hammelburg, Freising and Aachen, also introduced cost covering feed-in schemes for PV systems in their area of responsibility in 1993.

The introduction of a federal law on a technology-dependent FiT with a respective level to allow a positive return on investment, oriented by the ideas of Wolf von Fabeck and the experiences already described, was the big breakthrough in 2000, substantially driven by Hans-Josef Fell from the Green Party, Hermann Scheer from the Social Democrats and many more supporters at the time and thereafter.

5.4   Feed-in Tariffs – Sustainable Versus Boom and Bust Market Growth

Until 2007 Germany was by far the biggest PV market. This changed in 2008 after Spain introduced a similar feed-in tariff in 2007 which boosted

Table 5.4 Different Feed-in tariff's in 2008 for Germany and Spain.

Germany (20 years)	<30 kW	– 100 kW	>= 1 MW
€ct/kWh	46.75	44.48	43.99
Spain (25 years)		– 100 kW	– 10 MW
€ct/kWh		44.03	41.75

new installations to 2.7 GW, the largest market for PV installations in that year. In 2009 the Spanish market disappeared from the landscape with less than 100 MW of new installations – why?

The success story of the German feed-in tariff (EEG) was soon adopted by many countries in Europe and globally. More than 50 countries have by this point installed a similar support mechanism. The first country to follow was Spain, which introduced a tariff that was much too high compared with what existed in Germany. Table 5.4 summarizes the situation in the year 2008 for the two countries. It can be easily seen that for similar system sizes there was almost the same tariff with two important differences: firstly, solar radiation in Spain is much higher, by about 50%, and secondly, the tariff in Spain was decided to be paid not for 20 years, as in Germany, but for 25 years (and even for the years thereafter with a 20% tariff reduction). Since the major components, the modules and the inverters, are traded internationally, there is not a big price difference between the two countries. Even if the project development and installation in a new market is slightly more expensive it can be concluded that the generation cost in Spain was about one-third less than in Germany. The return on investment in Germany was aimed at around 8% for private systems. Due to the conditions just stated, investors in Spain could make a double digit return for more than 20 years, which naturally attracted many people and companies.

Another important difference between Spain and Germany was the origin of the feed-in tariff budget: in Germany it had been decided to spread the cost among electricity users (with some exemption rules for large power sensitive industries) leading to a small increase of the electricity price. This "small increase" is estimated as follows. At the end of 2008 we had ~5 GW of PV systems installed in Germany, which with the sunshine in this country produced (~1 kWh/W_{pv}) about 5 TWh. These TWh receive the above stated tariff of ~45 €ct/kWh, resulting in an annual cost of 2.2 billion € to be paid for 20 years. The value of these TWh at the stock exchange is not constant but varies according to the prevailing market conditions. If we take the average number of the past years which is about 5 €ct/kWh we obtain 0.25 billion €. The difference between paid feed-in tariff and stock exchange value – in our case 1.95 billion € - is called in Germany "annual Wälzungssumme" and the total amount in 20 years, if the stock exchange value were constant, would be 39 billion €, which would be the "total Wälzungssumme" or cumulative feed-in tariff budget. It is this number which is often used as a negative argument by opponents of such a support scheme. The real development until today and anticipated in the future will be discussed later.

If we take the annual budget for the year 2008 when we had ~5 GW cumulative PV systems installed and divide it by the total electricity consumed in Germany (minus the exempted big industries) of ~500 TWh one obtains 2.2 billion € / 500 TWh = 0.44 €ct/kWh. This is obviously a small addition to the German electricity price for households, which in 2008 was about 23 €ct/kWh (relative 1.9%). It should also be noted that the annual budget in real currency goes down by the inflation rate of 2% to 3 % per year, hence in 20 years the 2.2 billion € are only 1.3 billion € or 0.27 €ct/kWh which may be compared with the electricity price in 20 years of about 42 €ct/kWh, assuming a 3 % price increase per year (relative 0.6%). The "burden of a mortgage" as it is often called, is substantially decreasing in future years in real terms compared to today's financial burden. This simple calculation assumes a constant electricity consumption in Germany. All in all there is a wide acceptance among the general public for this sort of support. It is the retroactive measures of different kinds which make the situation unpleasant because they create unforeseeable risk, which makes financing more difficult, at least more expensive: additional taxes introduced in the Czech Republic, additional exemptions for industrial companies in Germany and unfortunately other retroactive changes in Europe.

With the exceptional PV installations in Germany in the years 2010 to 2012 with ~7.5 GW each year the cumulative installed PV power increased to ~33 GW at the end of 2012, which is a factor of ~6.5 compared to 2008. Due to the digression of the FiT the budget for the additional installations did not grow proportionally but with only by a factor 2.5 towards ~5.4 billion €. With the same calculation as for 2012 we have at the end of 2012 an additional 1.1 €ct/kWh for PV which is a total of ~1.5 €ct/kWh for all PV installations. For a household in Germany with an electricity consumption of 3,500 kWh this means a monthly addition of ~4.4 € to the electricity bill (for the total FiT including all renewable technologies this is 15 €). When listening to the agitated discussion in politics where in particular PV is blamed for "unaffordable electricity prices" I wonder whether one pack of cigarettes or preferably four scoops of ice cream per month are no longer affordable. The goal of this negative campaign is, for me, the ultimate attempt to have at least a quick and dramatic decrease in additional unwanted decentralized PV systems in Germany. A recent study by Prognos [5-2] concluded that with the current compensation in the EEG law an increase of the share of PV electricity from 4% in 2012 to 6.8% in 2016 – which is a 70% increase – would add less than half a cent per kWh – which is only 1.9%.

In contrast, the situation in Spain is one where most of the budget had to be paid by a ministry from tax money. As everyone knows, budgets are scrutinized every year and unforeseen expenses are very much disliked. And this is exactly what happened there: after introducing the generous tariff in 2007 it took many months to get approval due to bureaucratic hurdles. Hence, the minister did not notice how many applications were in preparation and when the green light was given for so many systems, in 2008 the minister could only watch as his budget shrank with each additional MW of installed PV systems. He did what he had to do in this situation in order to regain control of his budget and decided on a fixed cap of 500 MW for 2009 and thereafter- an upper bound that was never reached because investors found themselves in an insecure situation. The lesson learned was that, for a sustainable and long-lasting support scheme one cannot rely on budgets from ministries. It should also be mentioned that the tariff which was too generous also provoked a number of "black sheep" who were tempted to misuse this support, e.g., asking for unreasonable prices for necessary grid interconnection points – so it is important to keep the return within a reasonable range and not to overdo it. A similar boom and bust market occurred in the Czech Republic in the year 2010 with 1.5 GW and then a hard stop with almost nothing in 2011.

In 2009 it was Germany who once again led the new installations with 3.8 GW and the big increase in 2010 was in large part due to the 7.4 GW installed in Germany. Fortunately, new markets in Europe and worldwide also started to add significant installations, notably the European countries Italy, the Czech Republic (but see above the described bust in 2011), Belgium and Spain as well as the non-European countries the US, Japan, China and Australia.

When comparing installation numbers in different sources one should take a close look at the definition used. At the EPIA (European Photovoltaic Industry Association) we used the PV installations in a given year which were grid connected and monitored by official bodies in the respective countries. This gives the most trustworthy number for a country. However, especially in times of end-of-year-rallies, a lot of systems are installed in the last quarter in order to receive the tariff from the running

year before the next digression starts (mostly in January), the numbers
for "grid-connected" and "installed" PV systems can vary considerably.
For example in 2010 there were two countries, Italy and France, which in
addition to the grid-connected systems installed about 3.5 and 1.35 GW,
respectively, which could not be connected to the grid for time reasons. In
our EPIA market report for 2010, a global market number of 16.6 GW and
27.6 GW for 2011 was given. In other studies counting installed systems
you find higher numbers for 2010 and lower ones for 2011, which can easily be understood by just looking at the installed systems for these two
countries. For installed systems this gives a number of 16.6 + 4.8 = 21.4
GW for 2010 and 27.6 – 4.8 = 22.8 GW for 2011, which is comparable with
many other studies discussing installed systems. The two approaches also
have a severe consequence: the "grid connected" approach shows a huge
growth of 66% from 2010 to 2011, with Italy as the biggest contributor in
2011 while the "PV installed" approach shows only a moderate 6% growth
with Germany as the biggest contributor. It should be mentioned that in
the numbers above, I assume that no significant installations in 2011 were
made without also connecting them to the grid in the same year. A word
of caution: always pay attention to which methodology is being used when
comparing market numbers from different sources.

It is also interesting to see the cumulative PV installations in the main
geographic regions as summarized in Figure 5.6. Due to the large growth
in 2010, it can be seen that in this year alone, the volume of PV installations was almost the same as over the last 30 years. At the end of 2011 we
had about 67 GW PV installed and in late 2012 we surpassed the 100 GW
level globally. Let us compare this with the utility scale power producing
technologies in two directions, power-wise and energy-wise.

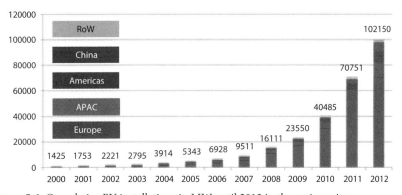

Figure 5.6 Cumulative PV installations in MW until 2012 in the main regions
(source: EPIA).

On a sunny day the power of all PV installations is around 85 GW – the variance from the 102 GW number is accounted for by the power at standard test conditions (these are measured at 25°C and printed on the data sheet) minus the difference to the actual power given by the respective solar radiation (direct and diffuse) in the field (for the temperature influence on efficiency see the later chapter on technology). These 85 GW are equivalent to the power output of 65 full-size nuclear 1.3 GW reactors! The energy production in a year is also quite remarkable. If we cautiously assume a 1.3 kWh/W_{pv}[2] number for all systems globally installed we obtain 110 TWh electricity produced which is equivalent to the energy output of 12 full-size nuclear 1.3 GW reactors (assuming 7,000 hours per year running time). In this context, one may compare the planning and approval time with the time needed to construct and commission just one nuclear reactor – 10 years is very optimistic. In only 5 years (2008-2012), Germany installed some 27 GW PV with an energy production of about 3 nuclear reactors. Once the question of energy storage has been economically solved– which again is just another technology development and also boosted by e-mobility – the portfolio of renewables will outpace the traditional energy technologies in a surprisingly short period of time.

In only 10 years, from 2000 to 2010, the manufacturing landscape completely changed. When comparing Table 5.3 (Top 10 in 2000) with Table 5.5 (Top 10 in 2010) one can see a remarkable development. In 2010 the spread for the production volume in absolute numbers was between 2 and 40 MW, while 10 years later it was a staggering 600 to 1600 MW. Other important observations can be made for this same time interval. One is the fraction of "the other" manufacturers, which increased from about 20% in 2000 to 50% in 2010. Another is the change in regional distribution: From the four Japanese Top 10 in 2000, Sharp invested most quickly and gained a 30% market share in 2004 and 2005 when the runner-up only had a market share of about 10%. A strong increase in production was accomplished by German newcomers, like Q-Cells and Solar World. For example Q-Cells, which was founded in 1999 and produced its first solar cells in 2001, went straight to be No.1 in 2007, overtaking Sharp and followed by the first big Chinese manufacturer Suntech. Due to a major support scheme in China to conquer the world production for solar modules by offering billions of $ of investment money (more than 30 billion alone for capacity increase)

---

[2] A PV system installed in a northern country like Germany produces 1,000 kWh/kW_{pv} or 1 kWh/W_{pv} and in regions like southern Italy this number is almost 2 kWh/W_{pv}; hence the assumed average should be conservative.

Table 5.5 Top 10 solar cell producers in 2010.

Company	% share	MW shipped
Suntech (China)	7.9	1572
JA Solar (China)	7.3	1464
First Solar (US)	7.1	1411
Yingli Solar (China)	5.3	1062
Trina Solar (China)	5.3	1057
Motech Solar (Taiwan)	4.6	924
Q-Cells (Germany)	4.5	907
Gintech (Taiwan)	4.2	827
Sharp (Japan)	3.9	774
Canadian Solar (Canada/China)	2.9	588
Others	~47	~9,400
Total	100	~20,000

the number of Chinese newcomers increased their production in a short period of time. As seen in Table 5.5 the regional distribution in 2010 comprised five Chinese companies, two Taiwanese, and only one each for US, EU and Japan.

The further development until 2012 is shown in Table 5.6, again with major changes in such a short period of time. First the total volume increased by 50% and the share of the Top 10 decreased to ~40%. As the numbers for the companies below Top 6 become quite similar there is also some uncertainty. For example the report by IHS based on production numbers no longer features Sharp but lists the Singapore based company Flextronics in its Top10. The future of Suntech is also open, since they have filed for bankruptcy.

It is astonishing to see that China which has been the No.1 in solar thermal collector production and installation for some time, is making a major move to also become No.1 in solar and wind. In other words, China is preparing to lead the most promising growth industries for the coming decades. Within the running 5 year plan (2011 – 2015) which has a budget of about $1.3 trillion they are not only willing but also able to make this happen. The next step will be to grow the Chinese market for PV installations in a similar way to what they achieved with wind installations. It should be no surprise that China will be No.1 in new PV installations within the next years and with ~13 GW new installations in 2013 they have been the country with most PV systems installed.

The balance in production volume of the single highest value item of a PV system, the module, and the respective market size in the various worldwide regions is shown in Figure 5.7 for the time between 2000 and 2010. In 2000 there was a good balance with only the US producing significantly more than its market share. The increase of the EU market

Table 5.6 Top 10 module manufacturers in 2012 (source: Solarbuzz).

Company	% share	GW shipped	Company	% share	GW shipped
Yingli Green Energy	6.3	2.0	JA Solar	2.8	0.9
First Solar	5.9	1.9	Jinko	2.8	0.9
Trina Solar	5.3	1.7	Sun Power	2.8	0.9
Canadian Solar	4.7	1.5	Hanwha SolarOne	2.5	0.8
Suntech	4.7	1.5	RoW	59.1	18.9
Sharp Solar	3.1	1.0	Total	100.0	32.0

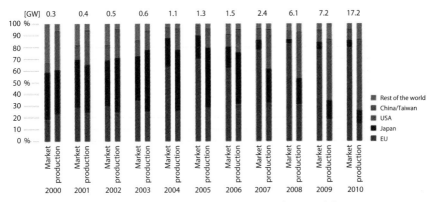

Figure 5.7 Market allocation versus regional production of PV modules.

share from slightly less than 20% in 2000 to 80% in 2010 did not result in a similar growth in module production, which in fact declined from slightly above 20% in 2000 to about 15% in 2010. The big winners were China and Taiwan, growing their production share from nil in 2000 and to well above 50% in 2010. Japan is particularly interesting, as it managed to more or less keep the balance between market and module production share. It should already be highlighted here that this comparison does not show the whole picture. As will be described in a later chapter the real value added in a national economy is much higher than can be anticipated from the picture above, because the imported modules from China and Taiwan contain a number of value added components created in Europe, e.g. materials (poly silicon, screen printing pastes, encapsulation materials) and production equipment.

5.5   PV Market Development Towards 2020

It is always a challenge to make projections of the market development for any industry – it is even more difficult to do so for an industry which has been driven by various support programs which cannot simply be extrapolated to the future in a business-as-usual mode. This was already described in the sections before for some countries. The strong growth in recent years meant that we grew almost "unnoticeably" to a size where all of a sudden considerable business was taken away from the traditional utility sector. This will certainly result in a reaction and in Germany and also Italy, the two biggest deployment countries, there is a severe argument pushed by the traditional energy sector and supported by biased politicians that the growth of PV installations should be stopped and that for example in Germany they

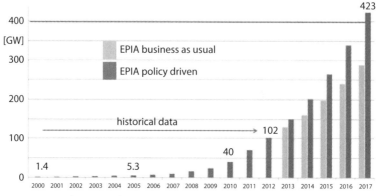

Figure 5.8 10 year history and two 5 year bottom up forecasts of the cumulative PV market: low growth (yellow) = moderate (Business as usual) and high growth (red) = policy driven (source: EPIA "Global Market Outlook 2013-2017").

should even be cut back from 7.5 GW down to 3 GW and less for the rest of this decade. This is a shame and a scandal and definitely fails to realize the great potential of decentralized electricity production by PV in combination with low voltage smart grids for the future. On the other hand we have the new countries which have just started to realize the positive impact of PV on their future power supply. Important examples are China, India and also the US as well as some South American countries and South Africa.

In this context, a bottom-up approach is difficult but still useful if a range is defined within which each of the regional markets could develop. At EPIA, a useful time horizon of 5 years is taken and repeated in a rolling year to year approach. Thereafter, we will also consider a top-down approach for a slightly longer time period until 2020.

5.5.1   Bottom-up Epia Market Forecast (2013 to 2017)

Every spring, our EPIA association has its annual general meeting where most of the members convene. This is a very good opportunity to take an additional day to discuss the global market of the preceding year and also to make projections for the coming five years for all relevant regional markets. As our members come from all over the world they can provide an in depth on-site contribution from their respective countries. The EPIA office sends out a questionnaire a couple of weeks before the assembly, asking each member to provide their best estimate for all the markets they have good knowledge of under two assumptions. The first which is called the "moderate" scenario should only take moderate political support schemes into account. The second, called the "policy driven" scenario assumes the best respective support scheme in all countries being considered. The big question remains what impact an assumed support scheme will ultimately have on market size. In Figure 5.8 the result of the EPIA market workshop from March 2013 is shown. As can be seen, the expectation in the moderate case was even a possible decrease of the global market and in the policy driven scenario a potential increase to 47 GW for the then running year 2013. A realistic range for 2013 could be around 37 GW. Also for the later years a much lower growth of between 9% and 22 % is anticipated compared to the average of 51% per year for the decade before.

5.5.2 Top-down Epia Market Forecast ("Set for 2020")

The SET plan – Strategic Energy Technologies – was adopted by the European Union in 2008 with two major timelines:

The long term goal is set for 2050 and targeted at limiting climate change to a global temperature rise of no more than 2°C, in particular by reducing EU greenhouse gas emissions by 80% – 95%.

For 2020, this plan provides the basis for the development and deployment of low carbon technologies in order to reach the well-known 20-20-20 goals by 2020: 20 % reduction of CO_2 emissions (even 30% in reduction is foreseen under some boundary conditions), a 20 % share of renewables in the end energy and 20 % increase of energy efficiency.

In a study by the European Commission it was found that with the current trend we are not in line with either the reduction of CO_2 or the renewable portion for end energy. The gap that was found is quite substantial. With business as usual, the predicted CO_2 reduction was merely at only 15% – far away from the 20% or even 30%. The 20% end energy, which amounts to 3,200 TWh (a total of 16,000 TWh was expected in Europe (EU 27) as end energy in 2020)

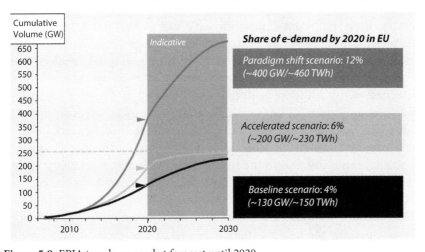

Figure 5.9 EPIA top-down market forecast until 2020.

was found for a business as usual case to be only at about 2,000 TWh – so there is a large gap of 1,200 TWh which has to be closed. Photovoltaics at this point in time was generally not believed to be able to contribute a major share to electricity production. Therefore, EPIA launched a study "SET for 2020" (in analogy to the more general SET plan described before) together with the consultancy company A.T. Kearney to analyze in a top-down approach three growth scenarios and their potential impact on closing the above gap.

When the study was carried out during the autumn of 2008 and summer of 2009 we knew that the global market in 2008 which was 6.2 GW with Europe making up about 5 GW; the cumulative installations in Europe at the end of 2008 were 10 GW. Taking a look at the envisaged electricity consumption in 2020 we obtained the figure of 3.850 TWh from the EU Commission (corresponding to an increase of 20% over 2010 electricity consumption). This brings us to the three following scenarios: (a) "baseline", providing 4%, (b) "accelerated" with 6% and (c) "paradigm shift" aiming at 12% of the EU's electricity produced by PV (see Figure 5.9). In order to reach the necessary 150, 230 and 460 TWh for the three scenarios, a cumulative installation of 130, 200 and 400 GW would be needed. In order to see what global growth would be necessary to reach these European numbers, we made the simplified assumption that the market share of Europe would continuously decrease from about 80% in 2008 to 50% in 2020. The resulting annual global PV market was then calculated to be 90, 110 and 160 GW for the three cases. The required growth rates between 2010 and 2020 would then have been 26, 28 and 34% per year, assuming a 9 GW global market in 2010. When looking at Figure 5.8 it can be easily seen that even the "paradigm shift" scenario should be do-able. But reality makes it even better. As outlined earlier, the actual installations in 2010 were about 21 GW which decreases the annual growth to the very moderate levels of 16%, 18% and 23%. An important finding is the significant contribution PV can make to closing the above mentioned gap of renewable TWh: with the paradigm shift scenario the 460 TWh would contribute more than a third to the needed 1,200 TWh.

Why did we call the high growth scenario a "paradigm shift"? During the course of our study we analyzed in more detail the ten most important countries within the EU 27 (to which we added Turkey and Norway) with respect to their potential cumulative PV installations and contribution of the needed TWh. Obviously this has to be connected to the specific country electricity needs and also to the irradiation conditions, as in our study we did not consider the transport of PV produced electricity from one country to the other. The six most dominant countries are summarized in Table 5.7 which account for 71% of cumulative installations and 74% of produced TWh. Especially the last line, the %-PV penetration of the country's total

Table 5.7 PV deployment in focus countries under the paradigm shift scenario.

Country	Germany	France	Italy	Spain	United Kingdom	Turkey
2020 PV Installed Capacity [GW]	80	60	55	40	22	20
2020 PV Produced Electricity [TWh]	79	73	78	62	21	30
% - PV penetration of total electricity consumption	13	14	18	18	5	16

annual electricity consumption, becomes an important parameter when adding more and more GW of PV in a country. For Germany, this is impressively demonstrated in a simulation done by the former institute ISET, now IWES in Kassel (Germany) and shown in Figure 5.10. The upper envelope of this graph shows the load curve in Germany for a typical summer week (here June) starting on Monday (day 1) through Saturday/Sunday (days 6 and 7). During the week there is a total load of between 40 and 60 GW with a pronounced peak load in the respective afternoons. Weekends typically show a reduction of the load to between 30 and 50 GW. At the time the simulation was done, we had 17 GW of wind power installed in Germany which for the particular June week was added as one contributor to the load curve. With the assumption of a homogeneous distribution of PV systems throughout Germany incremental steps of 10 GW PV systems were added and their output according to the actual insulation data of the June week considered can be seen in the graph. A first important fact is that PV obviously adds very efficiently to the afternoon peak load and reduces the otherwise needed power from peak load power stations. Up to about 30 to 40 GW PV installations, just the peak load is taken away. However, when adding additional GW's this power digs first into the medium power and then also into the base load power generation from utilities. The priority access of renewables to the grid causes additional challenges because traditional fossil and nuclear power stations have to reduce their power output when renewable power is available; because this is rather difficult with nuclear and coal fired power stations, utilities who run those units prefer

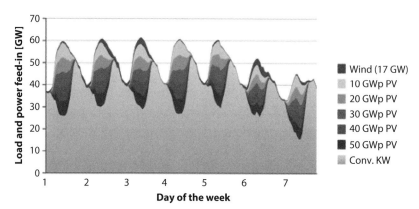

Figure 5.10 Electricity coverage of increasing PV installations in Germany from the country wide electricity load for a typical summer week.

to sell their electricity to neighboring countries for free, sometimes even to pay for accepting those 'too many' kWh's of electricity. Taking the 80 GW needed in Germany for the paradigm shift scenario and comparing this to the 50 GW shown, it can be clearly seen that for the Sunday afternoon the 100% load would be covered by PV alone – and all other power stations in Germany would have to be shut down in these few hours. Another important finding is the rise of a new peak load in the early evening hours – this is the first indication that additional measures must be taken when adding more and more renewables. Just imagine if we had 200 GW PV in Germany alone: most of the energy would have to be stored as the load in Germany would not absorb the actual power delivered. These additional measures are described in the next section and because of this we called this scenario a "paradigm shift" as a lot more has to happen besides installing additional GW's of PV. The main reason why in our study we allowed for a higher fraction of PV produced electricity in southern countries compared to more northern ones is the fact that in these countries the maximum load not only occurs in the afternoon – as is the case in the northern countries – but also occurs in summer, when most TWh of PV electricity are produced (Northern countries typically show a higher load in winter times).

5.6   Total Budget for Feed-in Tariff Support as Positive Investment for National Economies and Merit Order Effects for Electricity Customers

There is no doubt about the positive and outstanding impact of the feed-in tariff as the most effective support mechanism for renewable market development. This very positive element has of course been challenged by liberal economists, traditional utility companies and politicians with the major argument of it being too costly and a burden on the national economy. This section describes the opposite, namely that if all the budgets are seen as an investment for the respective country then, even if we apply conservative assumptions, we arrive at surprisingly a high Return on the Investment or a positive Net Present Value for this investment.

A first analysis of the additional value of a PV produced kWh compared to a gas fired peak power station (CCGT) was made in a study by ISET (now IWES) and published in the annual symposium in Staffelstein [5-3]. The annual report by LBBW "Valuing the invaluable" in 2008 [5-4] also analyzed which value in addition to just being another kWh electricity PV can offer, for example by introducing the concept of a "hedging value" for PV. Building on these results a major effort was undertaken within the previously mentioned EPIA study "SET for 2020". The basic ideas were as follows:

- PV systems do not emit CO_2, which is one value to be positively considered in relation to fossil power plants. As PV systems are able to replace peak power electricity production, a comparison with gas fired CCGT peak power plants was made. The CO_2 emission for these plants have been calculated as ~330 g/kWh and assuming a future average market price for CO_2 certificates of €38 per ton of CO_2 this results in a saving of €ct1.2/kWh by using PV (for oil and hard coal power plants this number increases proportionally to their increased emissions). If other market prices for CO_2 are considered this number changes proportionally. Although in Europe in 2013 CO_2 certificates are traded at less than €5/t-CO_2 this is unnaturally low for two reasons: first, a large number of such certificates was given to the industry for free when this trading started and secondly there is less industrial value creation in Europe due to the economic crises. Once this distortion is overcome we may return to certificate prices which really help cut down CO_2 emissions.
- PV can be treated as a hedge against increasing electricity prices. Once PV plants are installed there is only little running and maintenance cost for more than 25 years (which is the warranty many PV module producers are giving on their product). This implies that an investor can reliably assume that his generation cost for electricity over this time period will remain constant. This is very different in the case of traditional electricity production. In the gas fired power

stations running in Europe we currently use about 40% imported raw materials. This number is forecast to rise to 70% in 2020 (and most probably even higher in later years). As the gas price has a high share for CCGT produced electricity any change in highly volatile oil and gas prices will have a pronounced effect. In the LBBW study two hedging values of €ct1.5 and €ct3.1/kWh were discussed. The higher value would be consistent with an oil price scenario of $85/barrel in 2009, $110/barrel in 2012 and $135/barrel in 2018. The low value is consistent with 50 $/barrel in 2009, 63 $/barrel in 2012 and $100/barrel in 2018 (side note: when the study was carried out in 2008/09 one could have imagined the low price scenario, today in 2013 everything indicates towards the high price scenario). There is an even easier way to understand the magnitude of these two hedging values: a market study from the utility sector [5-5] shows a long term development in traded electricity prices between 2014 and 2034 of ~90 to ~€120/MWh for peak electricity (3,120 hours/year) and ~65 to ~€85/MW for base load electricity (8,760 hours/year), respectively. The price difference from the starting point in 2014 continuously increases towards 2 and €ct3ct/kWh over the 20 years, which equals a constant price difference of 1 and €ct1.5/kWh, which is the hedging value for the base load and peak electricity case mentioned above. If in future we also have a liberalized utility market in Europe, we will experience an even higher price for afternoon peak hours like in the US and Japan, where the above described increase over 20 years may well be in the range of 40 to €60/MWh, corresponding to a hedging value of 2 to €ct3/kWh. In summary, it is realistic to assume a hedging value for PV electricity in the range of 1.5 (low case) and €ct3.1/kWh (high case).

- In the EPIA study some other parameters were analyzed, like reduced grid losses (+€ct0.5/kWh), but also factors which reduce the value of a PV kWh like operating reserve (-€ct1.0/kWh) and lost margins for utilities (-€ct0.6/kWh).

Adding up all the above described individual value numbers in €ct/kWh gives:

(reduced CO_2 1.2) + (reduced grid losses 0.5) + (hedging value low case 1.5 or high case 3.1) – (operating reserve 1.0) – (utility lost margins 0.6) = (added PV value 1.6 for the low or 3.2 for the high hedging value)

We now did what every investor in the industry is doing: taking an investment, including all financial costs associated, and summing up all the revenues generated in later years by the investment. All the annual benefits are then discounted and a Net Present Value (NPV) calculated (major assumptions were a real discount rate of 3% and an increase for electricity prices of 2% between 2010 and 2020 and, very conservatively, constant thereafter). Only if this number is positive the investment pays off. In addition, if more than one potential investment can be done then the one with the highest Net Present Value should be executed as this one is the most rewarding.

In our case the investment is the total budget for the FiT needed to reach a certain growth scenario – the EPIA study looked at the three cases already discussed: base case, accelerated case and paradigm shift – and the benefit is the annually produced electricity multiplied by the "added PV value" described before. It is no surprise that for the three growth scenarios in Europe we had three different total FiT budgets: €155bn, €182bn and €235bn for the baseline, accelerated and paradigm shift scenario, respectively, as shown in Table 5.8. The calculated Net Present Value is shown for the three growth scenarios using the two hedging values in Table 5.8. A first important observation is that for all six calculations there is a positive NPV, even if small for the low growth and low hedging value. It is surprising to many that the NPV becomes significantly higher for the higher investments in the accelerated and paradigm shift scenarios. But when we remember that we have a decrease in the PV generation cost in later years due to the yearly digression and much more TWh produced for the higher growth scenarios this can also be understood qualitatively. Consequently, for a society it is not a burden to pay the early investors a feed-in tariff to stimulate volume but is in fact a clear positive investment. When seen over time, those PV systems produce clean and ever cheaper electricity.

As renewables (until recently mainly wind, but increasingly now also PV) increasingly enter into the area of trading electricity, there is an interesting phenomenon called "Merit Order Effect" which makes traded electricity cheaper because of the input of prioritized electricity from

Table 5.8 FiT budget and Net Present Value (NPV) for three growth scenarios.

Growth scenario	Baseline	Accelerated	Paradigm shift
FiT budget (bn €)	155	182	235
NPV (bn €) (low hedging value 1.5 €ct/kWh)	5	44	191
NPV (bn €) (high hedging value 3.1 €ct/kWh)	39	95	291

renewables. An example to explain this: taking a look back at Figure 5.10 shows that on a sunny weekday the traded electricity in the afternoon is typically at a high price because peak power stations have to provide this peak without renewables. If, however, PV is providing this peak electricity then the more expensive peak power stations don't have to be switched on and the traded price goes down. There have been a few times in Germany with strong wind conditions when the traded price even became negative. In general it is always the most expensive power stations which are shut down when renewable power comes onto the grid. This is why renewables have already shown an impact in reducing prices for all customers, even for those energy intensive companies who are exempted from paying the few cents per kWh which are used to pay the annual budget of the Feed-in tariff – which is obviously unfair to the many private households paying for this effect. With an increasing share of PV and wind power we have to redefine the value of those produced kWh's which will become a challenging task.

5.7   New Electricity Market Design for Increasing Numbers of Variable Renewable Energy Systems

Today (2013) we are in a critical and highly sensitive situation. Looking at the example of Germany, it can be summarized as follows: First, we have the political goal to stop the EEG when cumulative PV installations will have reached 52 GW. Depending on the annual installations this may happen in about 2017 (+/- 1 year). Second, the increasing volume of PV and wind will cause a further significant decrease in the price of traded electricity (due to Merit Order effect); lowest prices which may even be negative will be seen at times when wind and/or PV are supplying power equal to or above the load curve.

If we do not change our current electricity market landscape we will face the challenge of how to encourage self-consumption, i.e. consumption of electricity generated with decentralized renewable technologies like PV independently of the grid, and how to absorb electricity at meaningful prices into the grid when it cannot be used at the producer's site. Several aspects have to be considered after the expiration of the EEG:

- Continuing the existing EU directive on priorizing grid access for all renewables
- Self-produced electricity should not be taxed, at least for private households - otherwise one could also start asking for a taxation of vegetables grown in the own private garden
- The value of a PV kWh both for the owner of the PV system and society is at its highest when it is self-consumed: for the owner

because he compares his retail price with the LCOE (Levelized Cost of Electricity) of his PV system (including storage, DSM (Demand Side Management) and IT (Information Technology) to boost self-consumption) and for society because those kWh do not use the grid infrastructure and services

- The direct sale of self produced electricity should be allowed to neighbors within the Low Voltage Grid area, like decided by a new law in Italy early 2014
- For all electricity from PV systems that is not self-consumed we have to develop a suitable model for calculating a fair value of those kWh's with municipalities and other electricity providers, DSO's and TSO's (Distribution System Operators and Transmission System Operators). We should not work with price finding mechanisms from the past but should find new models based on a future renewable Energy world – it cannot be that those who invest in PV systems (and other renewable technologies) receive (almost) nothing for their kWh's because an outdated stock exchange mechanism is used which was developed for other framework conditions.

It will take a major effort to make this change happen. As we have to significantly change well established patterns, all stakeholders have to be involved with political moderation – a lot will depend on having the "right and forward looking" political guidance. One important aspect: should the traditional energy sector not make substantial efforts for the fair integration of renewables with suitable market mechanisms it may force many home owners and SMEs to quickly shift to 100% autonomy. Should this prove to be financially successful it may cause the floodgates to open.

5.8 Developments for the Future Energy Infrastructure

5.8.1 Smart Grids as the Future Low Voltage Grid, Distribution and Super Grids

The existing electricity grid – everywhere – originates from a time when big central power stations produced base and medium load electricity with some peak power stations closer to the users. There was only one way for the electric power, namely from the big power station via the very high AC network (360 – 500 kV) first to the distribution grid (20 – 50 kV) and from there to the local low voltage grid (380 V) from where it was finally directed to the individual customers, households or small businesses. The houses were normally equipped with a meter which still in most cases electromechanically counts the electricity used. This technology has now been in use for many decades and does not contain any intelligence. The described situation will change dramatically in the coming years and has started in a few countries with first steps. For example, in the late 1990s Italy introduced the first smart meters in houses. In existing liberalized electricity markets the need to also charge households with different tariffs is already in use for example in Japan and the US. This will also be the case in future, gradually in all regions. It is the low voltage grid which until recently could be seen as "a stupid grid" as there was no possibility to actively take part in the distribution of electricity but only to react to the needs of the consumers. This will have to change in the future if we seriously move to a renewably powered world. Here we have to collect electricity produced by all renewables in the best possible way and make an optimized portfolio of the various renewable technologies – which will be different in future for the various countries. So far, customers were mere consumers of electricity, but with a PV system or other generators in the house they will also become electricity producers; sometimes this future customer is called a "pro-sumer".

Figure 5.11 summarizes the situation of the various grid levels. We start with the green circles symbolizing the low voltage grids in a country. In the lower row this is shown for an urban situation, where a cluster of several low voltage grids is in close proximity (a) and for a rural case, where two situations are sketched for a larger (b) and smaller (c) grid area. It will be this low voltage grid which will develop and become the smart grid of the

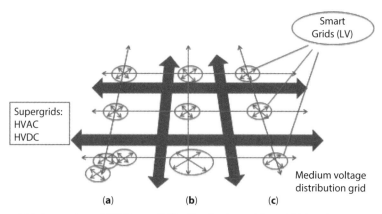

Figure 5.11 Low voltage (LV) Smart Grids, distribution grids and high voltage Super Grids.

future. While "smart" today is unfortunately overused, in this context it nicely describes a variety of intelligent ways which were not in use until now.

On top of the medium voltage distribution grid the notion of "super grid" is often discussed. The best known are the 360 kV to 500 kV AC transmission lines which have the disadvantage that they have high losses over larger distances. Although there are still possibilities to enlarge the transported electricity with such HV-AC grids (for example by allowing higher temperatures in the cable) this is limited. Thanks to the developments over the last two decades with high voltage DC grids, this is no longer an issue: with a 500 kV DC the losses are only about 3% per thousand km of electricity transported. This makes them ideal for trans-national electricity transport and to integrate them into future off-shore wind-parks. Whether this trans-national super grid could evolve into a world-wide super grid (which will be discussed in the last section) remains to be seen and depends mainly on the cost development for storage systems in all different sizes and storage times.

5.8.2   Supporting Measures for the Future Energy Infrastructure

Just a few ideas on what a future smart grid with intelligent meters and local production of renewable electricity will offer:

- The electrical appliances of households are closely monitored and controlled by the intelligent meters and optimize
  the self-consumption of locally produced electricity. The
  weather forecast for the coming 24/48 hours is also integrated to steer some of the appliances in times of highly
  probable electricity production from PV systems and adjacent wind power systems.
- In parallel to the well-known AC circuit, future houses may
  well have an additional DC grid as many appliances already
  run with DC power today (for example lighting, communication equipment and more). Households will also be
  equipped with a DC battery bank of 10+ kWh storage capacity to shift some of the afternoon produced PV kWh's to the
  evening. The batteries used here must only fulfill the lowest storage cost for electricity; weight is not an issue for this
  application. Hence the well-known lead acid battery may
  still have a long life of usage ahead of it. A medium time goal
  is a storage cost of 5 to 8 €ct/kWh. Future inverters will also

actively take part in the service of the low voltage grid, for example by providing needed reactive power. In addition, an optimized balance between DC power from the house-integrated PV/battery system into the parallel DC circuit of future houses, and feed-in of the unused kWh into the local low voltage AC grid will also be done.

- The family electric car with its battery is also integrated into the optimized production and usage of electricity in the respective household. These batteries have to be optimized in two directions: they must not only be low cost per kWh stored but also have a high ratio of kWh per kg battery weight. This is the main reason for today's Li-ion battery systems in this sector.

- An interesting approach for future electric cars is the range extender offered with the Ampera from Opel (similar also in US by GM) and the brand-new BMW i3. These cars run purely on electricity with all the advantages that brings with it (for example future wheel hub motor). But when a longer distance has to be covered, a small turbine recharges the battery while driving. It could be an interesting option to consider this range extender (running with biofuel or gas produced by renewable electricity driven electrolysis to form hydrogen and by adding CO_2 to get CH_4) being used as a power and heat generator in the house during night and winter when needed. A quick calculation for Germany demonstrates how this idea could help overcome the often discussed challenges for power supply when no wind and no sun provides power: at a 20% share of electric cars (~10 million e-cars) where half are equipped with a range extender (average power 20 kW) this fleet could provide 100 GW power – which is more than the highest power requirement in Germany during the year. This would of course need an integration of the e-cars into the Smart-Grid and IT network.

- All produced kWh not used in the own house are offered to the neighboring households, SME's and offices within the low voltage grid area. A new business opportunity for municipalities will develop as they can efficiently store all locally produced renewable electricity not currently used in the local grid in a larger battery bank in the transformer stations which will also communicate with other adjacent smart grid areas. Battery sizes could be in the MWh+

range and use Na-S battery systems for example. Such high temperature systems are not useful in households but can be well managed in a professional environment and offer lower cost per kWh stored electricity – a goal of below €ct5/kWh is realistic. Other storage technologies for this use could be redox flow type batteries which due to their costly infrastructure per unit (pumps and storage tanks for the electrolyte) have the advantage of larger capacities and due to their very low discharge they are well suited for storing electricity over longer periods of time.

These are just a few ideas and some food for thought when it comes to the future of smart homes, electro mobility, smart grids and intelligent interaction with local municipalities. The latter can effectively serve as a connection to the multitude of other municipalities via an exchange of information and electricity. Multidirectional transport of electricity within the smart grid areas is done with the medium voltage distribution grid. Input also occurs through wind parks and larger PV parks.

6

PV Value Chain and Technology

6.1 Basics of Solar Radiation and Conversion in PV Cells

The solar spectrum outside the earth's atmosphere has a power density of 1,300 W/m²; this is called the "solar constant". The energy distribution for the photons (energy per wavelength and area) is shown in Fig 6-1 and resembles the radiation of a black body radiator with a temperature of 6,000 Kelvin – the temperature at the surface of our sun. The visible part of the spectrum from about 400 nm (ultraviolet) up to about 700 nm (red) is only a small part of the complete spectrum. The relationship between energy, frequency and wavelength for a photon is described by

$$E = h \nu = h c / \lambda$$

with E the energy of the photon, h Planck's constant, ν the frequency, c the light velocity and λ the wavelength.

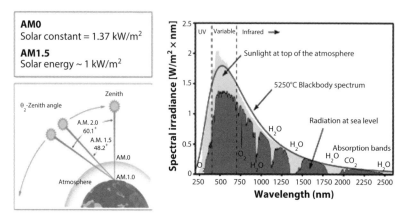

Figure 6.1 The solar spectrum above the earth's atmosphere and at sea level.

Entering the atmosphere, some absorption of light takes place due to water vapor and other molecules and aerosols in the air. The absorption bands at the various wavelengths of H_2O (900, 1,100, 1,400, 1,900 and 2,500nm), O_2 (750nm), O_3 (250nm), and CO_2 (2,050nm) are indicated and result in the decrease of the yellow area down to the red one. This decreases the power of the sunlight which is coming from the outside of the atmosphere. If the sun is in the zenith (perpendicular to the earth surface at noon), that is then the minimum path of the photons through the atmosphere and called "airmass 1" (AM 1, see also Figure 6.1 on the left). All other times, the photons travel a longer distance through the air making a pathway of 1.5 or 2 times the minimum distance (AM 1.5 and AM 2, respectively) which leads to the spectra also shown in Figure 6.1. The integral below the AM 1.5 spectrum which is taken as the standard spectrum for comparison purposes, is about 1,000 W/m².

The challenge now is converting most of the spectrum into useful electricity through PV solar cells. Typically solar cells consist of a semiconductor, characterized by a so-called band gap (E_g). The energy levels of the ensemble of electrons responsible for forming a crystal out of the individual atoms are characterized by the so-called valence band. If a photon possesses energy just equal to the band gap, this photon can be absorbed by the electron and kicked out of the binding site up into the so-called conduction band, where it can move freely in the semiconductor material. Is the energy of the photon smaller than E_g, this light particle passes through the semiconductor material without absorption. However, if the energy is larger than E_g, the excess energy is transformed into heat (thermalized) and cannot contribute to the electric power generation of the solar cell.

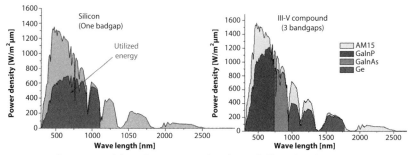

Figure 6.2 Absorption of the solar spectrum by silicon (left) and multi junction III-V semiconductor (right).

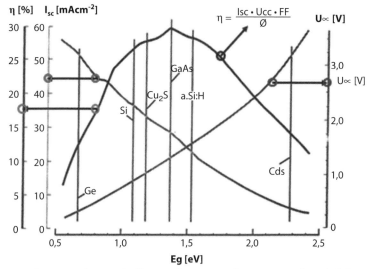

Figure 6.3 Current, voltage and efficiency for semiconductor materials with different band gaps (at AM 1.5 and 1,000 W/m²).

This has an important consequence as shown in Figure 6.2(left) in the case of silicon, with a band gap of 1.1 eV (corresponding to a wavelength of 1,100 nm). All photons with longer wavelengths are lost, because their energy is smaller than the band gap (grey part in picture 6-2 to the right of the red curve). All photons with greater energy than the band gap (grey part in picture 6-2 above the red curve) will lose that part of their energy which is above the band gap, as it is thermalized (lost as heat). Only the red part of the curve can be utilized as useful energy while the remaining grey part is lost when silicon is used as a semiconductor material.

Different semiconductor materials have different band gaps as shown in Figure 6.3. The energy output of a solar cell is given by the product of the short circuit current I_{sc} (proportional to the number of photons absorbed), the open circuit voltage U_{OC} (a function of the band gap of the semiconductor material) and a so-called fill factor (mostly dependent on losses caused by series and parallel resistances in the real device). If we take a material with a small band gap (small U_{OC}), most of the spectrum is absorbed (high I_{sc}).The other extreme, material with a high band gap (big U_{OC}) will only absorb a small part of the spectrum (low I_{sc}). The obtainable efficiency η (in %) is calculated as the quotient of the energy output of a solar cell (I_{sc} x U_{OC} x FF, where the Fill Factor FF is a device dependent constant) divided by the full solar power (AM 1.5) on the same cell area. The efficiency as function of the band gap is shown in Figure 6.3 where a clear maximum can be seen for semiconductor materials with a band gap of around 1.5 eV.

If we wish to extract more from the solar spectrum than in the silicon case, we have to use two or more materials with different band gaps, as is shown in Figure 6.2 (right) for a three band gap device. One can see that we not only absorb most of the spectrum and increase the open circuit voltage of the device (depending on the number of junctions used, here it is 3), but that there is also much less thermalization, as can be seen by the much smaller remaining grey part above the respective colored areas compared to the silicon case in 6-2 (left).

After the absorption of a photon by a binding electron and the electron being lifted up to the conduction band, there is now a vacancy in the valence band, also called a hole, which acts as a positive charge. If nothing else happens, the electron from the conduction band will fall back into the hole, thereby giving up the energy as heat or radiation. Therefore, we have to create something to separate the two opposite charges, the negative electron and the positive hole. The most appropriate method for this is an asymmetry built into the PV cell by doping the pure semiconductor material on one side positively and negatively on the other by replacing only few silicon atoms with other suitable ones.

Again, we use silicon as a demonstrator. If we replace some of the silicon atoms with boron, which has only three electrons in the outer shell, there are positive holes created in the valence band of the crystal (p-type conducting silicon). If we alternatively exchange some of the silicon atoms with phosphorous, which has five electrons, the result will be excess electrons (n-type conduction) via the conduction band. If we now dope a silicon crystal from both sides differently this results in an asymmetric transport structure: the two differently doped regimes of the crystal act (besides of the photon absorption) like selective transport membranes of the positive and negative charge carriers, respectively. This results in a collection of the photo-generated electrons at the n-doped side of the crystal and the collection of holes at the p-doped side. Thus a voltage is created between the opposite sides of the solar cell. The equilibrium energy of the charge carriers (electrons and holes) in undoped silicon, called E_f (Fermi energy), is located in the middle of the energy gap E_g, while for doped materials the Fermi energy moves nearer to the conduction band (n-doping) or to the valence band (p-doping), respectively (see Figure 6.4). If we bring the two doped semiconductor materials in contact with each other, the Fermi energy under no illumination must have the same level in the two doped materials as seen in the right side of Figure 6.4. It is this asymmetry which causes the separation of the light induced electron-hole pairs. There is – not intended for physicists –a simplified and descriptive picture for the flow of electrons to the contact material on one and holes to the other side of contact material of the solar cell. The flow of electrons can be seen like marbles rolling downwards, pulled by gravity, while the positive holes in contrast – like gas bubbles in water – follow the upwards gradient caused by the asymmetry.

Technology development is now all about finding the most suitable materials, cell architectures and manufacturing processes in order to very effectively fulfill and meet the physical boundary conditions described in a cost effective manner. The first solar cell was developed in 1954 by Chapin, Fuller and Pearson at the Bell Labs in the US [6-1], using silicon wafer and semiconductor know-how. Since then, we have seen a fantastic development from first cell architectures as seen in Figure 6.5 towards more sophisticated high efficiency ones which will be described later.

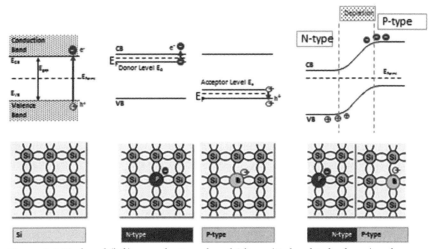

Figure 6.4 Undoped (left), n- and p-type doped Silicon (2nd and 3rd column) and pn-junction (right) under non illumination with associated band energy diagrams (top).

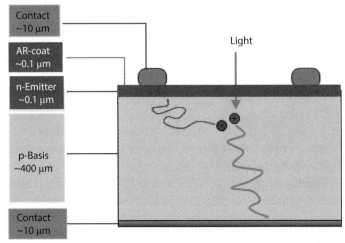

Figure 6.5 Cross section of an old type simple industrial silicon wafer based solar cell (thickness of the various constituents is indicated).

The basic cell design in the 70/80s was based on Czochralski wafers. The bulk crystal is normally p-doped with boron. A pn junction was formed in a diffusion furnace. During this process phosphorous accumulates within a thin layer of several hundred nm which then becomes negatively doped (overcompensation of the initial boron doping) and becomes the negative terminal of the solar cell. This thin layer is called emitter while the positive bulk material is called base and the rear side is the positive terminal. Contacts were formed as a grid on the front and as a whole area covering contact on the back. An Anti-reflective coating (AR-coat) was applied with a thickness of about 100nm.

6.2   Value Chain for Crystalline Silicon PV Systems

The first exercise will be to analyze the value adding steps to make a crystalline wafer based silicon solar module. Although PV modules are an important part of a complete system, there are additional crucial components and services which will be described later. We all expect a PV installation to last for more than 30 years and this in the harsh outside environment under (almost) all climatic conditions. This implies that not only the modules have to survive this long period of time, but that the mounting structures, interconnection boxes and cables have to as well. It is also important to highlight the workmanship required for mounting, as this also plays an important role in ensuring a long lasting and well operating PV system.

Depending on the purpose of the PV system, additional components are needed like DC-AC inverters for grid connected systems or charge controllers and batteries for stand-alone (off-grid) systems.

6.2.1   Poly Silicon

The semiconductor material required for this cell type is highly purified silicon. It is important to note that whatever the demand of solar PV will be in the future there will be no shortage of this material, as silicon is the second most abundant material in the earth's crust and available practically everywhere. As of today, we use quartz (SiO_2) as a starting material with only a few % impurities – we could ultimately also use sand by simply adding an additional cleaning step. This material is then reacted with coke to form metallurgical silicon, still with a few % impurities. Several million tons of this metallurgical silicon is produced per year and it is used as an additional material in many products, for example in steel making.

The metallurgical silicon is then reacted with hydrochloric acid to form chloro-silane. This gaseous material can then be purified by fractional distillation in large purification columns. Once the required level of purity is reached, the clean chloro-silane is decomposed in so called Siemens reactors where additional silicon is deposited on a silicon seed rod by CVD (chemical vapor deposition) at elevated temperatures of around 1,000°C. These rods are then crushed and used for further processing. Until recently, just a little over 10 years ago, only a few chemical companies like Hemlock and Wacker focused on the purification process for the semiconductor industry.

Only about 20,000 tons were needed for the semiconductor industry and the small amount of silicon needed for the PV industry until the turn of the century could easily be satisfied with the tops and tails of the mono-crystalline Czochralski crystals made for the electronic industry, and some other byproducts. At that time the market was about 200 MW per year and with 10 g/W the need was 2,000 tons, which made up just 10% of the existing poly Silicon capacity. This situation changed dramatically with the market growth as described previously. We already passed the GW mark in 2004, two years later the 2 GW mark, in 2008 installations reached 6 GW and in 2010 there were more than 20 GW of newly installed systems. Even by subtracting the Thin-Film products and using the decreasing needs towards 6 g/W today, it quickly became clear that poly silicon was the bottle neck of our industry. In the way that markets always react, we saw an increase in prices as well as new business models like prepayment, which requires the customer to get involved in the partial financing of the investment for additional poly Si capacity. In those days in addition to their co-financing of additional capacity, chemical companies could convince their customers to sign an annual "take or leave the promised quantity of poly Silicon, but pay anyway" contract. This situation has changed in the meanwhile and we are back to a more normal business situation, where each company in the value chain shoulders its own risks in the selling of future quantities of material, and bears the financial commitment for the necessary investments.

I remember that when I visited one of the big chemical companies in around 2002 and asked them to invest in more poly Silicon capacity quicker I was confronted with the question as to whether the PV market would really become a growing and big market. I answered honestly that with the German EEG I was expecting large growth. The only other major PV market at that time was Japan, which also supported the installations of PV systems politically. This, however, was taken by one of the board members as a major reason not to make an investment because: "the uncertainty associated with political support programs does not create an industry sector into which a respectable company would invest considerable money". If this company had followed my advice they would have been the first to enlarge their poly capacity and they would have earned hundreds of millions of Euro during the upswing of the PV market. I should, however, also add one important reason why the major chemical companies producing poly silicon at that time were hesitant to invest in more capacity: when the big semiconductor producing companies asked them to invest in additional poly capacity during the IT hype in the late 1990s, they did so and the downswing of the IT industry meant that there were no more orders for poly silicon than there had been before – so the chemical companies had a lot of idle capacity and substantial associated losses.

It is interesting to look at the price development of poly silicon over the years. At times when the PV industry could live with the waste of the semiconductor industry, the price was – very much depending on the quality – anywhere between 10 and $50/kg. During the transition time, when the chemical industry filled up the still existing overcapacity of the semiconductor market with dedicated solar poly silicon, the price for long term contracts was in the range of $60/kg. When silicon became the bottleneck of the PV industry, there was a quick increase of the spot market price towards $400/kg. This was often confused with the price for the majority of traded poly silicon. Bearing in mind that we used about 8 g/W in those days, the poly silicon alone would have contributed $3.2/W to the module cost where the finished product – including crystal, wafer, cell and module making – was sold at about $4-5/W. The losses would have been much too high, if the industry had used this spot material for the major part of their product! Spot market volumes were only used in special circumstances.

Today we have many more companies who invested heavily in more poly silicon capacity, notably Chinese companies and one South Korean company (OCI Ltd.). Just as an example, the Chinese utility company GCL Poly Energy Holdings Ltd., which started poly silicon production in 2007, ramped their production up to 1,850/ 7,454/ 17,800 metric tons in the years 2008, 2009 and 2010, respectively. I have never seen such an aggressive scale-up for a production which is known to require immense chemical know-how. Also, their reported production cost of $20/kg is quite remarkable. Figure 6.6 shows a very useful graph about the production companies in 2011 together with their estimated production cost and capacity taken from GTM Research [6-2]. This shows impressively that with the known needs for poly Silicon for the semiconductor industry (~30,000 tons) and PV industry (~180,000 tons) those companies with a production cost above ~$25 to 30/kg will have substantial difficulties to compete in the market. As a matter of fact there are a number of smaller companies exiting this industry. The first three bars show that new technologies like FBR (fluidized bed) and UMG (upgraded metallurgical silicon) have a slightly lower cost compared to the state of the art Siemens process. Another important message from this graph is that the four biggest companies (Wacker, Hemlock, OCI and GLC) could provide ~150,000 t of poly silicon, which is ~70 % of the global demand.

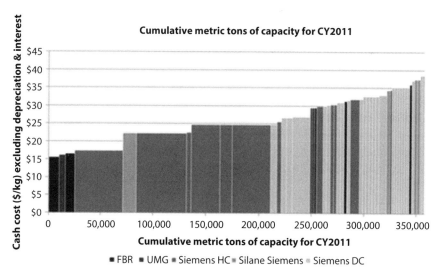

Figure 6.6 Production capacity and estimated cost for poly-silicon companies (source: GTM Research [6-2]).

During 2012 and also in 2013 there was no additional capacity building due to the above mentioned consolidation process. But even with a reduction of the 2011 capacities to levels around 250,000 t there is a surplus of silicon compared to the overall needs.

If anyone had asked the specialists only 10 years ago whether such a high capacity would be available today, you would have received a clear answer: "it will never be possible". The price levels of less than $30/kg and approaching below $20/kg in the spot market is also remarkable and is possible with an anticipated production cost of $20/kg and less – which is not necessarily the end of the game. All this would not have been seen only a few years ago.

Breakthrough versus mainstream: FBR versus Siemens reactor

Today's mainstream process to convert the purified silicon gas into microcrystalline material for further use is the so-called Siemens process described above. This is a very energy intensive process and a batch type one. Major efforts have been undertaken to develop a process using less energy. This was successfully done by the company MEMC in the US (renamed to SunEdison on May 2013 to reflect the company's focus on solar energy) which came up with a so-called "fluidized bed reactor (FBR)". Here, the purified silicon containing gas is introduced into a reactor containing small particles of silicon which grow bigger and bigger until they are so heavy that they follow gravity and can be extracted continuously as silicon particles of several millimeters in size. A similar FBR was developed at REC (Renewable Energy Corporation) and Wacker. Although the process at Wacker was already at the pilot stage it was not introduced as a new mainstream manufacturing process. As so often, when the process development started there was a good business case for the introduction of a new technology; but the continuous development of the mainstream technology often brings down production cost to levels which could not have been envisaged even by experts. This fact is impressively shown in Figure 6.6 where a production cost for a Siemens-based technology of below $20/ kg was perceived as being very challenging some years ago.

6.2.2   Crystal Making and Wafer Production

There are two different crystal materials in use: mono-crystalline and multi-crystalline. The first was and is in use in the semiconductor industry. In the so-called Czochralski process a big mono-crystal is pulled from a silicon melt with remarkable sizes today: up to 400 mm in diameter and with a length of more than a meter, weighing more than 400 kg, hanging on a tiny seed crystal. The development of mono-crystal diameter is a nice example of continuous technology development: starting in the early 1970s with a diameter of only 25 mm, it was enlarged to 50, 100, 150, 200, 300 and now 400 mm. While in the beginning of our industry we used round wafers like in semiconductor processing, today the crystal rods are either squared or pseudo-squared by cutting four slabs from the round crystal in order to have more cell area for a given later module size.

Alternatively poly silicon is melted in a big coated ceramic crucible and a multi-crystalline material is formed by controlled cooling. The size of these crystals is about 800×800 mm^2, they are 250 mm high and weigh about 400 kg. These big crystals are cut into 25 bricks of 156×156 mm^2 with slightly less than 250 mm height by wire saws.

It will be interesting to see how the future split between mono- and multi-crystalline wafers will develop from an almost equal share today. Only some years ago when poly silicon was the bottleneck of our industry all different grades were needed and this mixture of not always very pure material was well suited for the multi-crystalline process as cleaning takes place due to segregation of impurities during crystallization. As mono wafers have better electronic properties they are well suited to produce solar cells with high efficiencies of above 20%. Although they do need a clean and well purified poly silicon material, this is no longer a problem since with the built-up capacity for new poly silicon materials we have material which is good enough to go this route. My personal preference goes towards increasing the share of mono wafers and high efficiency solar cells and modules for the crystalline silicon technology.

An interesting new approach was recently introduced and described in a patent by BP [6-3]. Using the crucibles for multi-crystalline ingot making together with a new seeding process at the beginning of the crystallization, it is possible to grow a silicon block with a mono-crystalline content as high as 80%. This could lead to the marriage between the multi- and the mono-crystalline worlds, which provide high productivity large volume crystals and high electronic material quality for high cell efficiencies, respectively. A number of companies (e.g. JA Solar, Yingli, GCL, Trina, LDK) offered this material with different product names to the market (e.g. Virtus-, maple-, quasi-mono- or mono-like-wafer). However, even in 2013 companies did not switch mainstream to this new process as the remaining multi-crystalline fraction as well as partially extended defect formation in the mono area itself, causes a number of quality problems which are not yet solved [6-4].

Both materials, the squared mono-crystals and the multi-crystalline bricks are placed under a mesh of high speed moving wire in a water/

glycol bath containing abrasive particles (SiC, Al_2O_3). The spacing between the wires in the mesh defines the future wafer thickness while the wire thickness causes a considerable loss of silicon material, also called kerf loss. Very slowly – it takes several hours – the mesh moves through the crystal and the wire drives the abrasive particles, thereby grinding (or lapping) through the respective crystals. As an alternative to the – quite costly – process using abrasive materials, diamond coated wires are also in use. At the end of the cutting process the wafers are cleaned. Today's wafer thickness is about 180 µm with a kerf loss of about 120 µm. This is another remarkable result of continuous technology development. 30 years ago, wafers were cut by diamond coated inside hole saws which caused a kerf loss of almost half a millimeter with a wafer thickness in a similar range. Hence, in the old days one used almost 1 mm of crystal for one wafer which can be compared to 0.3 mm today – more than a threefold reduction! As crystallization is a highly energy intensive process it is not surprising that the so-called energy payback time was quite high in the early days. In the 1990s, the first wire saws were introduced driven by the PV industry but at first with thicker wires and thicker wafers. The better material utilization also convinced the semiconductor industry and today this is the state of the art of wafer cutting for them as well.

Breakthrough versus mainstream: ribbon versus wafer slicing

Especially in the early times like in the 1980s there was an obvious need to develop alternative processes for making wafers which would not waste good silicon crystal material into kerf loss. Mobil Solar started a development in the 1970s by pulling silicon sheets out of a melt without any kerf loss. The silicon sheets had a thickness corresponding to the required wafer thickness. This so-called EFG (edge defined film fed growth) process used a graphite crucible with a capillary in the form of a closed multifaced ring. Molten silicon, which is wetting the graphite material, creeps up the capillary and can be attached to a seed silicon sheet which can be pulled upwards with a velocity of a cm per minute. In order to get a better productivity per pulling machine, the Mobil experts very soon made a closed circumference, first with 9 sheets of 25mm, then 50 mm and later an octagon which was 100mm in width and 350nm in thickness. The length of the pulled silicon tubes was an astonishing 5 to 6 m. The machines and the tubes can be nicely seen in Figure 6.7 together with three laser cutting stations, which cut the respective 100×100 or 125×125mm² wafers from the tube with only very small losses. The residual corners can be reused in the pullers after controlled crushing.

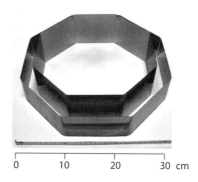

0   10   20   30 cm

Figure 6.7 EFG pullers and three laser cutters in the front with silicon tubes hanging (left) and part of two octagons with 100 and 125 mm (outer ring) side width (right) (source: SCHOTT Solar).

After the acquisition of the shares of Mobil Solar in 1994 and integrating them into ASE (Applied Solar Energy) as our subsidiary ASE Americas, a major effort was made to industrialize this process. By doing so we also experienced what it takes to develop a breakthrough process such as EFG in competition with the mainstream industry making crystals and sawing them. While we had to develop all the machines and processes within one company, the mainstream benefitted from a much larger number of companies, thereby dividing the associated development cost. While in the beginning the benefit of EFG wafers was substantial in terms of silicon saving by more than a factor of two, it melted down to savings of less than 1.5. Once the downstream processes for cell and module making are able to work with low breakage with 100μm wafers, which has already been demonstrated by wire saw cutting, this benefit will become even less. Hence the lesson learned is: if you are not able to become mainstream in an industry in a rather short time with a breakthrough process step, you had better stop this development; at least you should have a substantial cost saving with your new process compared to the mainstream process which should last at least for the depreciation time of the respective machine park. The EFG process was only recently stopped, in 2011. A similar ribbon process was developed by Evergreen in the US in the last 20 years and is called "string ribbon". The process was still in operation at the German company Sovello but the company went into insolvency in summer 2012. There is still one R&D activity on-going for a special ribbon wafer technology, called Ribbon Growth on Substrate (RGS). This was originally developed at BAYER Solar, transferred to Solar World and later to the Dutch research institute ECN. A pilot production machine was built at RGS Development BV, a Dutch development company. The potential cost advantage area wise for RGS-wafers cannot reach today's cost per Watt level compared to multi- and mono-crystalline silicon wafers because of the quality gap existing today [6-5]. This becomes even worse when going to the module level and taking into account the area related BOS costs.

6.2.3   Crystalline Silicon Wafer Based Solar Cells

This process step is crucial for determining the conversion efficiency of the complete system. Efficiency is an important lever for all area dependent value added steps. It starts with the price per area of a wafer. A cell process with 30% more efficiency (for example 21% instead of 16%) can afford, for the same price per Watt, a 30 % higher price for the same wafer. This higher level of efficiency again decreases the module making cost, all area related BOS (balance of systems) cost and the installation cost by the same percentage. It is therefore very important to always consider the complete picture for the installed system.

We will now take a closer look at the development of the process steps for making solar cells over the years. The first commercial solar cells were fabricated for the space application, first in the US and in the 1960s also in Germany by AEG. Those cells used evaporated metal contacts (Ti, Pd and Ag) and SiO_2 as antireflective coating. Efficiencies were in the range 10% to 12% and the cell size was $2\times2cm^2$ at first, later $2\times4cm^2$, and they were cut from round mono-crystalline wafers.

In the 1970s the industry started with polished round mono wafers similar to those used in the semiconductor industry. Until recently almost all wafers were p-doped and received a phosphorous diffusion on the front side in a batch process, where the wafers sit in a quartz boat moving into a tube furnace. After this process, the diffused region is not only on the front side but every-where, including around the circumference of the wafer. In order to prevent short circuits from the front p-doped area to the back side, either a mechanical grinding of the edges or later a plasma process was performed. Metallization on the front side was done through evaporation of Ti-Pd-Ag contacts either through a shadow mask or by chemical etching of most of the full area depos-ited metal, whereas back metallization was typically done with evaporated sil-ver. The antireflective coating with TiO_2 in the earlier days was done by an atmospheric CVD process. Typical efficiencies in production in those early days were around 10% for mono-crystalline and less for multi-crystalline wafers with a cost per Watt almost two orders of magnitude higher than today.

In order to cut costs, the expensive evaporation of noble metals was replaced by screen printing of the front grid with a silver paste and the back contact with an Aluminum paste together with silver dots or lines in places where the later series connection with soldered Nickel coated copper connectors is done. These screen printed pastes then have to be sintered to remove the organic components and for proper contact formation. In addition the sintered Aluminum paste forms a back surface field at the back side of the solar cell, which hinders recombination at the metal interface, thereby increasing the cell efficiency. Another important step to increase the efficiency was the replacement of the antireflective coating TiO_2 with an atmospheric pressure CVD process by SiN_x with a plasma enhanced CVD process. The hydrogen atoms from the plasma very effectively passivate not only the surface of the solar cell but also the defects in the bulk of the crystal, especially in multi-crystalline material, thereby boosting the efficiency of solar cells.

Breakthrough versus mainstream: MIS-inversion layer versus pn-junction solar cells

When I started to conduct joint research with Prof. Hezel in the mid-1980s (first at the University of Erlangen, later he was the director of ISFH) on a new type of crystalline solar cells, there were good reasons to do so. His patented MIS (metal-insulator-semiconductor)-inversion layer solar cell had exciting features compared to standard pn-type solar cells:

- elimination of the high temperature n-doping diffusion step
- exchange of material costly silver paste metallization with evaporated aluminum
- decrease of production steps for cell manufacturing

The challenges of transferring the university 2×2 cm^2 laboratory cell size into a 100 cm^2 industrial cell size (at this time) were the following:

- well controlled formation of a tunnel oxide with a thickness of about one nm
- evaporation of Al front contact through a shadow mask
- control and prevention of the potentially possible light induced degradation of the inversion layer by new UV blocking encapsulation materials
- interconnection of the evaporated Al-contacts for string formation by ultrasonic welding

All technical challenges could be solved with a 1 MW pilot line including large area module production with a fluid encapsulation technology. When looking into the next step towards a 10MW production line we had to realize that at this time there was no evaporation machine available from the equipment manufacturers which would have been able to make cost competitive front and back contacts. A major reason was the relatively low deposition rate which contributed to a machine cost that was too high for the contact formation. Today, 20 years later, the situation would have looked different as we will see in the later chapter for future PV developments. The lesson to be learned from this exercise is to not only to pay attention to the development of all necessary process steps but also to make sure that production machines are available from the equipment manufacturers for the required up-scaling of manufacturing capacity. In order not to discourage researchers from working on breakthrough technologies I will highlight two positive examples hereafter.

State of the art cell technology in 2013

The continuous development of crystalline silicon solar cells resulted in a modest but significant increase in efficiency of multi- and mono-crystalline silicon wafers from 8% and 10% in the 1970s to 15% and 18% in ~2010, respectively. Two companies, Sun Power and Sanyo, each pushed for higher efficiency with a breakthrough cell concept: the first with an Interdigitated Back Contacted Cell (IBC) solar cell and the second with a heterojunction solar cell (HIT). The IBC cell architecture placed both contacts at the rear side of the cell, thereby eliminating the shadowing from the front contact grid and bus bars. The efficiencies were around 20% and today up to 23% are achieved in production. The drawback was and is a significant cost increase for the cell add-on manufacturing which, however, helps to decrease the module and BOS cost structure compared to state of the art solar cells. The HIT solar cell concept had efficiencies of around 18% in the past and has been able to demonstrate up to 21% today.

It is interesting to see that the further continuous development based on the state of the art has also recently been able to show efficiencies on standard Cz-wafers with 156×156 mm^2 wafer area up to 21% (with screen printed Ag front contacts) and 20.9% (with electroplated NiCu front contacts). The cell design was developed at SCHOTT Solar and is shown in Figure 6.8 [6-6].

Only 5 years ago even the experts would not have believed that a screen-printed solar cell based on standard wafers could be manufactured with such a high efficiency. In addition the demonstration of electroplated NiCu

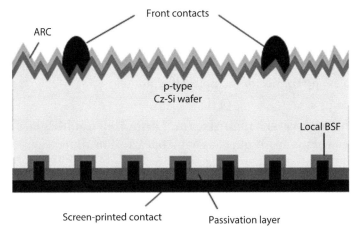

Figure 6.8 State of the art cell design by SCHOTT Solar.

front contact enables to abolish the expensive Ag-front contact paste. Unfortunately after these encouraging results SCHOTT Solar closed its PV business at the end of 2012.

6.2.4   Stringing and Module Making

Typical module sizes in the 1980s used 36 solar cells made of 100×100 mm² wafers and had a dimension of ~50×100 cm² (4 strings each made up of 9 series of connected solar cells) or 40×130 cm² (3 strings with 12 series of connected cells each) and a power output of 40 to 50 W. The number of series connected cells with an open circuit voltage of ~18 V was chosen to allow the charging of a standard lead-acid battery because stand-alone systems used to be the major application in those days.

Today's solar modules are the building block of much larger systems and their size has increased up to a level just to be handled by one person. These modules have mostly 10 cells series interconnected by tinned copper connectors, and 6 such strings are placed on the back-sheet plus encapsulation material of the later module and also series connected. The electrical data are about 30 V and 7 to 9 A and range in power from ~220 up to 280 W for multi- and mono-crystalline wafers, respectively. In order to have the highest power output one has to take care that all series connected cells have a very similar current at peak load, as by series connection the cell with the lowest current determines the total one. Therefore the cells have to be sorted in order to account for this which is done at the end of the cell making process when the quality is checked. After placing an additional layer of encapsulation material, mostly EVA (Ethylene Vinyl

Acetate), the front glass is placed on top and the total layer stack is heated under vacuum in an autoclave to fluidize the encapsulation material and interlink the constituents. Typical dimensions of framed modules are 1.7 × 1 m² with a weight of ~20 kg. A junction box with bypass diodes at the back with cables and connectors complete the module.

Breakthrough versus mainstream: Large area double glass resin modules versus front glass - EVA - back foil modules

When I was developing the aforementioned MIS-inversion layer solar cell at NUKEM in the 1980s I was also involved in the planning of a number of PV installations at RWE Energie. At the time of completion in 1988, the first one with a power of 340 kW in Kobern Gondorf was one of Europe's largest installations. The experiences from this test site were the basis for the second one of a similar size in Grevenbroich-Neurath near to a lake. One of the findings was that installation costs could be substantially decreased if the module size was bigger. This was the trigger for my R&D department to work on an increase in module size from the standard 0.5 m² to 2 m². In addition I wanted to make best use of our new cell technology which had the advantage of using much thinner wafers compared to the pn-cell technology. This was the trigger not to use the standard autoclave method with EVA but to use a fluid resin which polymerized either chemically or with UV light after filling the sandwich front glass – string matrix – back glass. This method did not require any pressure to be applied to the sandwich and was therefore best suited to work with much thinner wafers (less than 200 μm) compared to what was used as the standard (~400 μm) at the time. We could successfully demonstrate this module technology at the lake Neurath installation and also prove that there was a cost advantage in mounting these large area modules rather than the standard ones. However, it was the lack of standardized production equipment for this new type of module manufacturing compared to the mainstream which did not allow for a cost competitive production. A similar resin technology was developed at Pilkington Solar but was also stopped in favor of the standard module technology. It may well be that in future this resin technology may come back again when a cost effective silicone resin with high transparency and UV stability is available and a real need comes up to use wafers with ~100 μm thickness.

6.3   Value Chain for Thin-Film Technologies

A completely different way to produce solar modules is the so-called Thin-Film module technology. In contrast to the indirect semiconductor silicon, the absorber is characterized by direct band gap materials with high absorption coefficients which allow very thin layers to absorb the solar spectrum effectively. These thin layers cannot be handled separately as in the case of silicon wafers but need a substrate which can be a glass sheet or flexible steel or plastic foils onto which the absorber is deposited. There are two different methods of making Thin-Film modules, namely with substrate or superstrate technology. In the first case the substrate is used as a back-sheet and the various layers are deposited onto it. Module finish is done by laminating a piece of front glass to this stack of layers. In contrast to this, the superstrate technology requires the various layers to be deposited onto a front sheet. Module finish in this case can be the lamination of a plastic back sheet or of another glass sheet.

Figure 6.9 summarizes the production sequence for Thin-Film modules. For a superstrate technology like Thin-Film silicon (amorphous and microcrystalline) the sequence is as follows:

- Formation of a transparent conducting oxide by PVD (sputtering) process; the layer could be an indium-tin-oxide, doped Sn-oxide or Zn-oxide.
- Defining the individual 5 to 10mm wide cell strips by laser cutting, depending on open circuit voltage and current density of the device
- Deposition of the absorber material Thin-Film silicon or CdTe/CdS (Cadmium Telluride / Cadmium Sulfide).
- Parallel to the first laser cut a second laser cut is done
- Formation of back contact by PVD, followed by
- Third laser cut to finalize the interconnection of the various cell strips

In the case of substrate technology – e.g. for CI(G)S (Copper-Indium-Gallium-Selenide) – the production sequence is the exact opposite. It starts with PVD back contact formation, 1st laser cut, absorber material deposition, 2nd laser cut, PVD TCO front contact and 3rd laser cut.

As indicated in Figure 6.9 most of the production steps and used materials for substrate and module making are the same, or at least similar in cost per area. The essential difference is the choice of the absorber. Today's major Thin-Film technologies as summarized in the picture have different costs per area and result in devices with different efficiencies. The lowest cost per area is the close space sublimation for CdTe/CdS (CTS) modules which does not need a vacuum process. This is in contrast to the deposition technologies for Thin-Film silicon (PECVD) and CIGS (PVD), where vacuum processes contribute to a higher cost per area. With today's efficiencies the

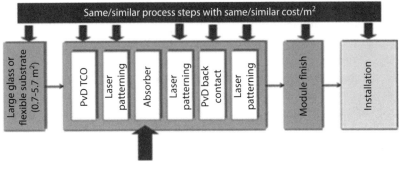

Figure 6.9 Process sequence for Thin-Film modules; efficiency ranges (as today in production → short term goals) for the various technologies in brackets.

lowest cost per Watt is obtained with CTS modules; in future it may be possible to reach higher efficiencies with CIGS compared to CTS (as indicated) which could result in a similar cost per Watt on a module level and an advantage for BOS and installation cost. Also from an environmental point of view it is advantageous to use a Cd-free technology which is possible for CIGS. The Thin-Film silicon technology, although best suited from a material availability point of view, suffers today from limited efficiency. Future chances are (i) higher efficiencies with multi band gap devices, (ii) less degradation (Stäbler-Wronsky), (iii) absorption enhancement to allow for thinner layers, (iv) increase in deposition rate and (v) combination with wafer based crystalline silicon technology.

It is important to understand the integrated series interconnection for Thin-Film modules which is one important difference to crystalline silicon modules. This allows all the layers to be deposited on large area substrates, which is one driver to decrease the specific cost per area. In addition, the series interconnection takes away the standard interconnect in strings of front and back contacted crystalline solar cells and all the potential difficulties attached to this. The schematics for this process are summarized in Figure 6.10 for Thin-Film silicon superstrate technology. The width w_d of the patterning by the three laser cuts are shown much larger than the cell width for reasons of clarity.

This patterning width does not contribute to the photocurrent and is also called inactive or dead area of the module. Obviously the ratio w_d/

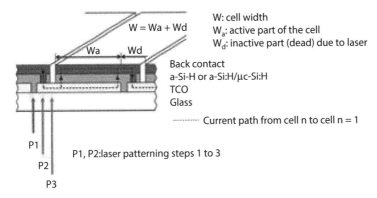

Figure 6.10 Integrated series interconnection with Thin-Film solar modules.

w_a should be as small as possible by maximising w_a, the active part of the module area, and minimizing w_d. The first objective is a trade-off due to Ohmic losses (voltage and current of the device and resistivity of the two contacts) while the second is determined by the patterning accuracy of the laser and the three laser cut widths. As these laser strips have to be accurately placed in relation to each other on quite large substrate areas – the largest ones today are 5.7 m² big – highest technical requests must be met, both mechanically and in terms of temperature control.

6.4   Concentrated PV (CPV) and III–V Compound Solar Cells

There has always been an attempt to utilize the cell area in a more efficient way. One of the first examples was the already described US project in the early 1980s, where the modules were mounted on a tracking device, following the sun during the day and a metallic mirror was mounted on both sides of the modules to reflect more light onto the modules. This was obviously only a very low concentration of a factor of about 2. In later years silicon solar cells were used in medium concentrating devices where lenses focused sunlight with a concentrating factor of about 10 to 20. Although there was the benefit of using less semiconductor material and a significantly improved power output over the course of a day (much more rectangular than the normal output of flat plate devices), these two axis trackers, however, significantly added to the balance of systems (BOS) cost of a PV system. This was the main reason why such concentrating devices did not make it onto the market place.

With the advent of the highly efficient III-V compound solar cells, developed for the powering of satellites, there is now a new business opportunity. The cell efficiency which one can buy today is about 40% and this will greatly lever the BOS cost for the tracking system. With systems that are in the field for a few years there will soon be a sound demonstration of the longer term performance and reliability of such systems, making them bankable for wider use. The claim today is that on a module level there is already a similar cost per power unit achieved compared to state of the art modules, and the additional tracking cost is compensated by the higher specific electricity output in kWh per kW installed compared to flat plate systems. Of course these systems can only be operated in regions with a high ratio of direct sunlight to global irradiation and will therefore nicely complement the use of flat plate PV systems. In particular, they might challenge the solar thermal concentrating systems for electricity production as discussed earlier.

In terms of technology, a typical 3-junction device is made up on germanium substrates onto which a complex layer stack of III-V compound materials with different band gaps together with tunnel junctions between the individual cells is formed by MOCVD (molecular organic chemical vapor deposition). As the three layer stack of cells is connected in series, close attention must be paid to match the current for all three cells, as the one with the lowest current will determine the current of the whole device. The cost per area for such devices is quite high but with a concentration ratio in the module of 500 up to 1,000 the specific cost per Watt becomes quite competitive. The typical dimensions for the module substructures are a plastic Fresnel lens with an area of $100 \times 100 \text{mm}^2$ which concentrates the sunlight onto cell sizes of $\sim 4.5 \times 4.5 \text{mm}^2$ or $\sim 3.2 \times 3.2 \text{mm}^2$ for concentration values of 500 or 1,000, respectively.

One module is typically made up of ~ 100 substructures and the panel placed onto the tracker is composed of ~ 100 such modules. The power output of such a panel is approximately 30 kW and even larger panels are currently in planning.

6.5   New Technologies (Dye, OPV, and Novel Concepts)

Mimicking the photosynthesis in plants has already inspired many scientists, among them Prof. Grätzel at the University of Lausanne (EPFL). I remember an interesting research project which I did at his institute in the early 1980s. At that time he was aiming for the ultimate: light absorption by a dye and splitting water into storable hydrogen and oxygen. It soon became clear that this task could not be optimized for the highest efficiency. A few years later he had the great idea to develop what is now

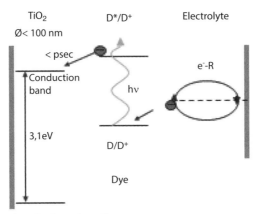

Figure 6.11 Principles of a dye solar cell.

called the "Grätzel cell". The basic idea was to create a layer of highly porous or nanostructured TiO_2 on glass, coated with a conductive layer. A light absorbing dye (Ruthenium-tris-bipyridile) was attached into the porous material. Due to the very high active surface (BET surface[3]) an efficient light absorption is possible. The important steps for harvesting the sunlight in these devices, which are fundamentally different to photovoltaic cells, are shown in Figure 6.11:

- **Light absorption by dye D**
  The dye D absorbs part of the solar spectrum and after light absorption an electron is kicked into an excited (triplet) state D^*.
- **Charge separation**
  Two possibilities now exist: either the electron falls back to its ground state where it came from and releases the absorbed energy as heat, or the electron transfer from the excited state to the conduction band of the semiconductor material TiO_2 is so fast that this becomes the dominant reaction. In the latter case the dye D^* becomes the oxidized D^+.
- **Useful work done**
  The electron is then transported via the conducting band of the TiO_2 to the outside electrical circuit and can perform useful work with the captured energy from the sunlight.

---

[3] The measurement of the specific surface area of a powder or porous material is interpreted following a model by Brunauer, Emmet and Teller (BET method). The BET surface is measured in m^2/g(of material)

- **Closing the circuit**
  This is done with an electron relay e⁻ - R (Methyle-viologen) which picks up the electron from a conducting electrode layer, brings it to the oxidized dye D⁺ and reduces it back to the original ground state D.

Maximum efficiencies with such devices in the laboratory have reached more than 10%. But from my point of view it is much more important for this cell type to use different dyes in order to create real colored modules. This is not possible with band gap semiconductors which are typically only grey or black. Although the efficiency for a colored module will be naturally smaller as only a fraction of the sunlight – depending on the desired color – will be absorbed, there will be customer demands for such specially colored modules, for example modules that are to be integrated into the housing of OEM (original equipment manufacturers) products or for special Building Integrated PV modules.

Another old dream is starting to become a reality, namely solar cells made from plastics – or even better: made entirely from organic materials. Good progress has been made in recent years with a number of big chemical companies involved. Major challenges for this cell type are:

- Still expensive materials, for example the often used C_{60}-cage molecules
- Long term stability – this may not be an issue for customers demanding only a few years of lifetime from their products

Beyond these new types of solar cells described above there are more fundamentally new ideas which are being worked on at many university laboratories. One example is band gap shifting in nanostructured layer stacks of otherwise well-known materials like silicon and silicon oxide. By alternating such nm-layers and changing the thickness of one of the layer constituents it is possible to shift the band gap of such a silicon layer stack from 1.1 eV to 1.3 eV, which may be interesting for future multi band gap cell structures. Another research topic is finding ways to utilize the energy of "hot" electrons, which means that when an electron is kicked above the lower conduction band with photons having an energy content larger than the band gap, this surplus energy should also be extracted and not wasted as heat. Another interesting research area is to utilize a two or more photon absorption process where the sum of the energy of these photons equals the energy of the band gap and both photons would be lost for contributing to the photocurrent where they would not normally be absorbed.

6.6   Other Cost Components for PV Systems

6.6.1   DC – AC Inverters

For grid connected PV systems the DC – AC inverter increasingly becomes the brain of the whole system. Not only must the DC current from the solar modules be converted with highest efficiency to grid compatible AC current under all conditions, but this device is also situated at the interface to the public grid and must increasingly contribute to grid service functions[4]. In addition this "little box" will also be the master switch and control for the "intelligent home" of the future. This home manager may then also communicate with the local weather forecast in order to be able to switch various appliances in the house according to the local renewable energy production – both the own production as well as the production in the neighborhood. In this way the self-consumption of locally produced PV electricity and other renewable sources within the local low voltage region can be optimized and reduce the use of the medium and high voltage grid network.

6.6.2   BOS – Balance of Systems

All hardware that is needed to complete a PV system besides the modules and the inverter is called Balance of Systems (BOS). Each system category – like roof-top, BIPV, green-field etc. – needs individual ways to mount the modules on a structure and to optimize the electrical wiring from the modules to the inverter and from there into the grid. The overall performance and also the lifetime of the system critically depends on these rather simple components, but if one compares the variety of specially tailored products in well-developed and emerging markets it can easily be seen that the mounting and wiring time is very different and this influences the associated price. In addition, auxiliary machines, for example to quickly and accurately mount hundreds of pylons in large area green field systems have significantly contributed to lowering the installation time. In order to further reduce the BOS material cost the voltage rating for the used modules becomes important especially for large ground mounted and commercial systems. Today's state of the art is 1,000V but higher voltages may help to further decrease this specific cost.

---

[4] examples are provision of reactive power and grid fault-ride-through capability in case of AC-grid difficulties

With optimized BOS components the installation time per kW could be drastically reduced compared to earlier times. The training of specialized craftsmen has also added to the quality and speed of installation. In Europe a major effort was undertaken to create the so-called "Solarteur" who incorporates the skills of an electrician (with emphasis on DC) and a slater (mainly for the roof PV systems). The availability of equipment to improve the safety of the installers was also an important accomplishment for the sector. While installation cost for a PV system was around 13,500 €/W in the early 1990s now, 20 years later, we have a price range of 1 – 1.5 €/W, depending on system size. Such a quick complete system price decrease for an electricity producing technology which was accompanied by a decrease in the levelized cost of PV produced electricity has never been seen before. It is the modularity of PV systems and the associated large volume manufacturing of the components which makes such a development possible.

6.7   Marimekko Plot for PV Systems and Summary Chart for Cell Efficiencies

Combining all the individual information from the complete value chain for an installed PV system can nicely be shown in a so-called Marimekko plot (Figure 6.12). This name not only stands for a Finnish textile and housewares company but also for a useful diagram to show the interdependence of the major value added steps for a PV system on the relative cost split for each of those value steps (sometimes also called a mosaic- or matrix plot). Along the x-direction the relative cost contribution for the various added values can be seen where the 100% stands for the installed system price. In the case of crystalline silicon this starts with the production and purification of metallurgical silicon of the required quality. The subsequent steps are crystal and wafer formation, cell production, module manufacturing, DC-AC inverter, BOS cost and installation. For each of the value added steps the y-axis shows the relative contribution for equipment depreciation, material cost, labor cost and overhead (for example electricity, rent etc.). The increase of Capex (capital expenditure) in the upstream area, manifested in equipment depreciation, and the increase in labor in the downstream area value added steps is clearly visible.

The major materials (also called OPEX = Operational Expenditure) for the value added steps are:

- **Purified silicon**
  metallurgical silicon, running material for the process (acids, gas etc)

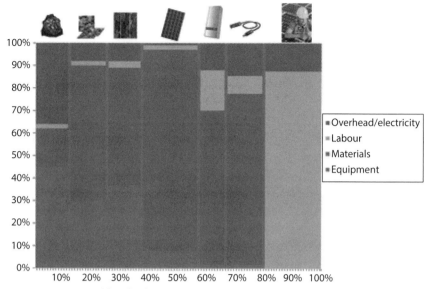

Figure 6.12 Marimekko plot for crystalline silicon PV systems (status 2011).

- **Crystal and wafer**
  crucibles, wire, slurry (for standard steel wires) or diamond coated wires
- **Solar cell**
  Al and Ag paste for front and back contacts, gases, dopants
- **Module**
  low iron and hardened front glass, encapsulation material (e.g. EVA), back sheet plastics or second glass sheet, junction boxes, cables and connectors

In the case of Thin-Film systems one cannot split the value added steps in making a module like in crystalline silicon. Instead, the absorbing thin layer, the cell device and the cell interconnect and module finish must be done in a single factory. We also have to take into account the generally lower efficiency of Thin-Film products by about 25% compared to crystalline silicon modules, which will increase the relative contribution of BOS and installation to the total cost and decrease the contribution for the module part. The corresponding Marimekko plot is shown in Figure 6.13.

A very nice summary chart on the status of all known efficiency data from the start until today was done at NREL (National Renewable Energy Laboratory in USA). It shows the interested reader how continuous development pushed efficiency up from the very beginning towards today's levels. Some important lessons can be taken from this chart in Figure 6.14:

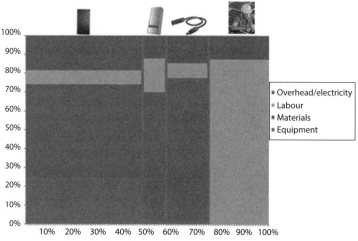

Figure 6.13 Marimekko plot for Thin-Film PV systems (status 2011).

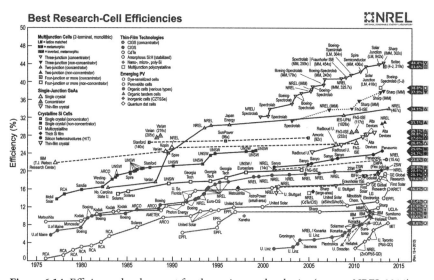

Figure 6.14 Efficiency development for the various technologies (source: NREL 2014).

- **Continuous improvement**
  for all technologies there is a steady increase in laboratory cell efficiencies
- **Leveling off**
  for single junction devices which have a theoretical efficiency limit it is obvious that as laboratory cells approach these limits we approach these asymptotically.

- **Continuous increase**
  only cell structures which still have room to reach ultimate theoretical efficiencies – like multi-junction devices especially in concentrators – still show a further efficiency increase. Just recently there was an announcement by SOITEC together with Fraunhofer-ISE, CEA Grenoble and HZB (Helmholtz center Berlin) who reported September 2013 a new world record with a 4-termional device with an efficiency of 44.7% under 297 suns (not shown in the graph).

This plot will serve as an important starting point when looking into the further development of the various technologies. It is also important to note that this efficiency chart does not indicate a direct link to the production cost of modules made from these technologies – but it may serve as a guideline for the important lever of efficiency and what it can do for the total system cost due to efficiency dependent area related cost components like BOS and installation. The chart also indicates at which university or company the respective world record efficiencies were reached. While 30 years ago the span of efficiencies was between 3% and 10% for Thin-Film modules, 16% for mono-crystalline silicon and 22% for GaAs cells we now have a much wider range, summarized in Table 6.1. It also indicates the efficiencies of commercially available products. Clearly there is a distinguished gap between the small cm^2-area laboratory cells and the typical $15.6 \times 15.6 \ (= 243) \ cm^2$ commercially produced Silicon solar cells and ~m^2 Thin-Film solar cells and modules. This area increase together with cost oriented production steps are the main reason for the efficiency decrease. Cell interconnection and the necessary space between the cells and the sealing edge around the module circumference contribute to the further decrease from cell to module efficiency.

Table 6.1 2013 status of world record cell and commercially available cell and module efficiencies (source: SERIS 2013 [6-7] and own data)

Efficiency [%]	CPV	cryst. Silicon		Thin-Film technologies				Other types	
	mj III-V	mono	multi	aSi/μc	CTS	CIGS	Dye	OPV
World record cell area ~cm²	45	25	20	14	19	20	11	10
Commercial cell area ~dm²	39	18–22	16–19	12	14	15	9	6
Commercial module	32	16–21	14–18	11	13	14	7	5

7

The Astonishing Predictive Power of Price Experience Curves

7.1 Basics about Price Experience Curves

The phrase "learning curve" was first used at the end of the 19th century by the German psychologist Hermann Ebbinghaus [7-1] when conducting experiments on how fast human beings acquire information. The first quantitative description is linked to an observation by T. Wright [7-2] from the Wright-Patterson Air Force Base in the US in the early 1930s that every time the total aircraft production doubled, the required labor time decreased by a constant percentage, which was in this case 10–15% (the learning factor).

The term "experience curve" was later used in a much broader scope. It states that each time the cumulative volume of produced goods or a service doubles, the value added costs fall by a constant and predictable percentage (the experience factor). If one plots (Figure 7.1) in a double logarithmic scale the number of units on the x-axis and the cost on the y-axis one obtains a straight line, where the slope represents the experience factor.

This was first described for single products or services in a single company. In the late 1960s Bruce Henderson of the Boston Consulting Group

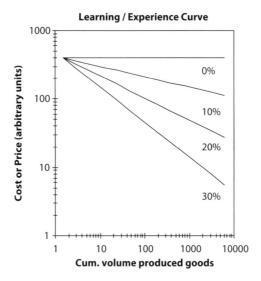

Figure 7.1 Principles of a Cost Learning- and Price Experience Curve.

(BCG) began to emphasize the implication of the experience curve for strategic considerations [7-3]. Research by BCG in the 1970s looked at experience factors for various industries, ranging from a few percent points up to 30%. One of BCG's proposals to consulted companies was that this predictable cost decrease should be passed on to the respective price development.

This graph is called a "price experience curve" (PEC). It was also found that for a specific product a straight line was not only obtained for a single company but also when one plots the global average price versus the globally produced cumulated number of units. Although there are many factors additionally influencing this PEC, it is remarkable that in most cases this straight line quite often follows several orders of magnitude.

PECs do not have time as a parameter, but only the cumulative volume of the product in consideration. This becomes important when extrapolating to the future. A price forecast for a future point in time can be derived by assuming a particular growth rate which results in a new cumulative volume.

7.2 Relevant Price Experience Curves Comparable to PV

An impressive PEC can be seen in Figure 7.2(a) for the DRAM semiconductor price per bit as a function of the cumulative bits produced over the last 30 years [7-4].

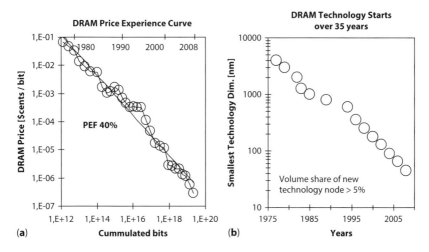

Figure 7.2 PEC for DRAM price per bit (a) and start of smallest technology dimension (b).

For more than 5 orders of magnitude, the price declined with a Price Experience Factor (PEF) of about 40% every time the cumulative volume doubled; this corresponds to an average price decline of about 33% per year. This behavior is technologically driven by the well-known Moore's law in the semiconductor industry (doubling of the number of transistors per unit area every 1.5 years). Figure 7.2(b) indicates the respective times, where a new so-called technology node was introduced. The latter is thereby described by the smallest technology dimension applied on the chip.

During my studies in the 1970s I remember that after the statement by Moore all scientists and industry technologists concluded that the PEC for silicon as a semiconductor material would come to an end in the 1990s at the latest, for material property reasons – no one could foresee that with photolithography it would be possible to create smaller and smaller dimensions from several microns down to today's 20 nm scale ("nano-manufacturing"). This is why today we still follow the same PEC with the same PEF for silicon –just with prices that are some orders of magnitude lower per bit. It can sometimes also help to calculate what a memory chip of 4 GB would cost today with the price per bit as it was in the 1970s which was just about a \$ct/bit at the time. Hence 4 GB would have a price of 4×10^9 bits x 1 \$ct/bit = 40 million \$(!) – which can be compared to the few dollars you actually pay for such a chip nowadays.

It is often argued that such a high PEF of about 40% as observed for the DRAM bits is only possible because of the miniaturization techniques applied in the semi-conductor industry. A remarkable PEC for flat panel displays which increased substantially in size over the last 20 years is shown in Figure 7.3(a) [7-5].

The need for Flat Panel Displays (FPD) began with the need to have them for mobile phones and laptop computers which could not be done with the CRT (cathode ray tube) technology. Starting in the early 1990s with a substrate size of only 0.1m², the technology for the TFT-LCDs (Thin-Film transistor - liquid crystal display) developed continuously towards 1.4m² (Generation 5) and, more recently, to more than 5m² (Generation 8). The technology for the next generation of 10m² has already been announced by display manufacturers. It is interesting to note that the PEF for this product is an astonishing 35%. This is due to the fact that most of the value-added steps profited significantly from the volume and size development. In Figure 7.3(b) the logarithm of substrate size is plotted against time. From the linear increase it can clearly be concluded that there is an exponential increase for the size as a function of time which in this case is the driving force for the PEC of this product.

Architectural glass deposition is also an impressive example for a long lasting PEC. The layer stack for a high quality low-e window is a complex one: for a two layer Ag system with a thickness of about 15 nm, approximately 10 individual layers (ZnO, TiO_2, Si_3N_4 plus special buffer layers) with thicknesses of between 2 and 20 nm are necessary – and all these layers have to be sputtered over the substrate size with an accuracy of better than 2%. If one plots the price for coating (determined by the price difference between of coated glass and bare glass per m²) over the cumulated volume of the millions m² of coated glass one again obtains a straight line with a PEF of 17%. This curve started with the first cumulated million m²

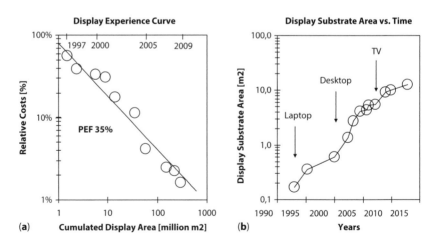

Figure 7.3 PEC for Flat Panel Displays (a) and start of new and larger substrate size (b).

in the early 1980s and in 2010 developed towards 10,000 million m². For standard float glass with a given thickness no major price reduction was experienced in the last years. This is due to the fact that the two major cost components for producing bare glass are material and energy, and prices for both have increased in the past. The PEF for a complete window glass system composed of one bare float glass and one coated glass has therefore decreased towards a value far below 10% [7-6]. The main reasons for the PEC have been the development of larger and larger glass substrates with a substrate size which today is the so-called jumbo size at 3.21 × 6 m². This will most probably not be further enlarged as there are size limitations placed by the transportation by road and railway from the float or rolled glass lines to the customers. The corresponding output of an architectural glass deposition line increased from 1,000s of m² in the beginning towards 10 million m² today; those production lines have a width of 4m and a length of more than 200m with a multitude of well controlled magnetrons to guarantee the layer thickness homogeneity described above. If one had asked 30 years ago when the first small area sputtered low-e depositions were made whether this would be possible on today's jumbo glass sizes most people from academia as well as from industry would have strongly opposed.

7.3   Lesson Learned from PECs Discussed

Qualitatively it can be stated that the higher the technology contribution and the lower the material part of the production cost of a product is, the higher the PEF; it decreases as the material and energy cost covers more and more of the total production cost of a product.

If for a given product the industry

- was already able to demonstrate a price reduction according to a product-specific PEC over the course of 1 decade
- R&D can provide further technology development for better device structures, cheaper and/or less materials and better processes

then it is highly probable that for this product a further price decrease will happen according to the same PEC with further volume increase. In many cases even technologists, be they from research institutes or from industry, could not and cannot imagine the further price development of their own product as predicted by the PEC.

Revolutionary changes for new products are very seldom: In the case of the PEC for the DRAM semiconductor, the natural start of the PEC was with the development of the first solid state transistor, which was a breakthrough technology replacing vacuum tubes. Since then, the basic transistor design has remained very much the same, despite the fact that many inventions have taken place. There will be a natural end to this type of product, when the dimensions simply begin to approach the atomic radius of the silicon material. Only then – after an astonishing product generation life time of about 70 (!) years – will we see a radical change in this type of product and the emergence of a revolutionary new one.

The same is true for the Flat Panel Display which replaced the old CRT (cathode ray tube). When it was introduced in the 1990s not many could foresee the technology development which make solutions available today which would not have been possible for the old CRT product. A possible replacement could be the OLED (organic light emitting diodes) which have brighter colors and less energy per area usage – provided that technology development can produce these structures from today's 30×30 cm^2 towards many square meters as for the FPD product. Product generation lifetime for FPDs will most probably be 30 years and more.

Photography is also a nice example of a long product generation lifetime: the good old camera with sophisticated reflex camera products was in use for more than 100 years. When the first solid state cameras appeared on the market in the 1990s the professionals did not foresee a quick introduction, especially not in the high end sector. But technology development made this happen and today solid state cameras are in everybody's use.

7.4   Price Experience Curve for PV Modules

7.4.1   Historical Development

The PEC for PV solar modules based on c-Si wafer technology is shown in Figure 7.4(a) with prices at the various cumulative quantities from the time such modules were sold in the market until today [7-7]. Credit should be given to Paul Maycock who – according to my memory – was the first to apply PEC for solar modules in his regular reports. For reasons of comparison the prices are taken for "power" modules only and also for modules based on crystalline silicon. It can be clearly seen that as in the examples described above we see a straight line with a corresponding PEF of 20%. Unlike for DRAM and FPD we do not have one single parameter driving the PEC, but a multitude of different parameters, summarized in Figure 7.4(b). Constituents for this reduction of the production cost were a reduction of specific material costs, increase of efficiency and "economy of scale" in production. This is exemplified by the specific Silicon material cost: the wafer thickness was reduced from 0.7 mm to 0.15 mm and the kerf loss from 0.5 mm (diamond inside hole saw) to 0.1 mm (wire saw); the material usage per dm^2 cold thereby be reduced from ~28g down to 6g. Together with the reduction of the price reduction in the chemical industry for poly Silicon from $60/kg to $20/kg the material cost per dm^2 was reduced from 168 to $ct12. The increase of efficiency from 8 up to ~20% resulted then in the specific poly Silicon material cost from $ct210/W in the 1980s to only $ct6/W today – this is a reduction by a factor 35! During this time the annual throughput of a single production line for cell and module production was increased from 1 MW (manufactory) to 200 MW (fully automated with high yield).

Sometimes the real price points deviate from the straight line which may have many different reasons. The increasing prices in recent years can be attributed to a shortage in the base material poly Silicon and to the market demand created in many countries. As the bottleneck has now been removed we not only quickly reached the PEC curve again but even see an under-shoot. This phenomenon is also known from the semiconductor industry with the so-called "pig cycles". Now, in 2013 at a cumulative volume of 100 GW there are prices of around $0.8/W. This is considerably below what one would expect from the PEC (~$1/W). There is – unfortunately – proven evidence that today we have unhealthy prices, proved by the fact that all major module production companies showed deep red figures in their 2012 annual reports. This is due to overcapacity in the producing industry and a slow-down of the market growth causing heavy competition and thereby resulting in the low prices observed. Interestingly enough is the fact that if one calculates backwards from the published annual reports from 2012, to find out at which price those companies would have had a reasonable positive margin at the volume sold the result is that just this $1/W which would have made the difference. Basically, technology development is driving the further development of the PEC as an overriding rule and the deviations from the straight line are caused by factors such as the one described above. It is worth mentioning that such deviations in both ways of real prices from the PEC are sometimes interpreted in a way to change the slope – or the PEF – for future development. While it may happen that in the future the PEF will gradually decrease when material and energy prices become the dominant part of the product cost (but remember the constant PEF for the DRAM and FPD), it is highly unlikely for the other way, i.e. increased PEF.

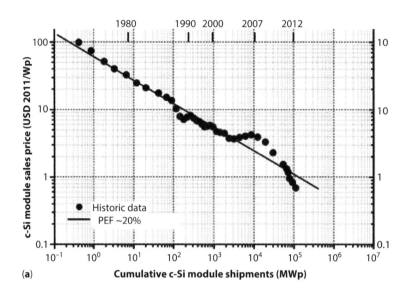

(a)

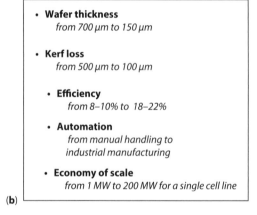

(b)

Figure 7.4 PEC for crystalline silicon solar modules; actual data with circles, straight line best fit (a) [7-7] and driving factors for this development (b).

7.4.2   Differentiated PEC for c-Si and Thin-Film Products

With Thin-Film power modules entering the PV market it was seen that for a given cumulative market size the corresponding price was significantly lower compared to c-Si modules. This is not surprising as the production technology is completely different in both cases as shown in Figure 7.5.

It will be interesting to analyze what the future price development for Thin-Film products will be under a set of assumptions summarized in

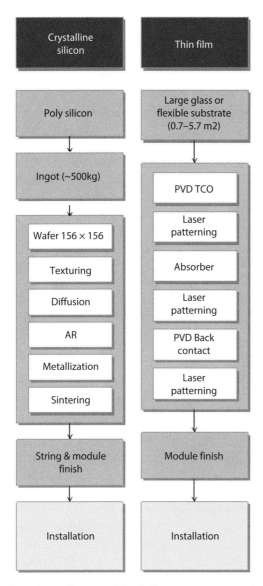

Figure 7.5 Value chain for c-Silicon modules (left) in comparison to Thin-Film modules.

Table 7.1. Based on the PEF for FPD products one is tempted to increase the PEF for TF products from 20% in the case for c-Si to somewhat higher numbers – together with my colleagues [7-8] we assumed a possible 25% in addition to the same PEF of 20%. We also looked at two scenarios with regard to the split between TF and c-Si modules in the time range towards 2020, which was either a constant number of 15% or increased linearly

Table 7.1 Set of assumptions for extrapolating the PEC for c-Si and TF.

PV volume growth scenario	Baseline				Paradigm shift			
Thin Film volume share	15%		15% → 35%		15%		15% → 35%	
Thin Film PEF	20%	25%	20%	25%	20%	25%	20%	25%
Case	1	2	3	4	5	6	7	8

towards 35% in 2020. The overall global growth rate was taken from a study by EPIA and used two scenarios as shown in Figure 7.6.

The Baseline scenario assumed a cumulative installed volume of 130 GW in Europe and for the whole world of about 280 GW, while the Paradigm shift scenario used 390 GW for Europe and 760 GW globally. When the study was conducted in 2009 we expected an annual market of about 10 GW for 2010 – which in reality approached 20 GW. This implies that the growth rate did not need to be as high as was then assumed, adding even more credibility to the overall picture. The first scenario would imply that in Europe we could produce 3% of the electricity which was expected to be needed in 2020 and the Paradigm Shift Scenario would even go up to cover 12% of Europe's electricity needs.

In Figure 7.7 the extrapolation of the future PEC for c-Si and TF is shown for the two extreme cases 1 (low growth, constant TF market share and same TF PEF) and 8 (high growth, increasing TF market share and higher TF PEF). For case 1 we expect a price in 2020 of about $ct 80/W for c-Si- and $ct70/W for Thin-Film modules while for case 8 we take from the PEC $ct60/W for c-Si- and $ct30/W for Thin-Film modules. Let us take a look at the assumptions to see whether they are realistic or too ambitious. First the two market growth scenarios: starting from 19 GW new installations in 2010 we need a very modest annual growth rate of 15% to reach the 80 GW assumed for case 1 in 2020. Remembering from the earlier market section that we saw an annual growth of 51% between 2000 and 2010, it is highly probable that the 15% are on the very conservative side. Even the case 8 where a market of 160 GW was assumed to be reached in 2020 we would only need an annual growth of 24% which is still far below what our industry was already able to demonstrate. Taking a look at the lower price number for Thin-Film modules in case 8 of $ct30/W it remains to be seen whether at prices below $ct40/W we will still follow the same slope or whether due to increasing material contributions to the total module cost we may see a gradual decrease in this slope, resulting in a price number slightly higher than anticipated. But these details do not change the

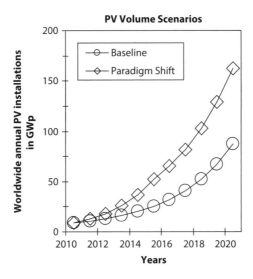

Figure 7.6 Future growth scenarios for annual PV installations until 2020.

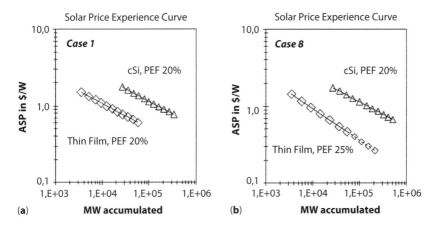

Figure 7.7 Extrapolation for the PEC of c-Si and TF modules for the two extreme cases 1 (a) and 8 (b) as given in Table 7.1.

general picture of continuously decreasing prices to levels where we can expect the LCOE (levelized cost of electricity) of PV electricity to decrease to numbers where conventional power plants will no longer be able to compete; this will happen earlier when fuel prices rise quicker and CSS (carbon sequestration and storage) will add even more to the overall cost of a traditional fossil fuel power plant. My simplified estimate taken from the sensitivity analysis is a more realistic price of $ct70 (+/-10)/W for c-Si- and $ct50 (+/- 20)/W for Thin-Film modules in 2020. It should again be highlighted that the given price levels would imply healthy prices, meaning reasonable positive margins for all companies along the whole value chain for PV modules, including component, material and equipment producers. Prices for the two module types that are much higher or lower than the given ranges are highly unrealistic in the time frame envisaged.

7.4.3   PV Systems Analysis

When looking to the lower price numbers in $/W of Thin-Film modules one is tempted to conclude that these products would soon dominate the total market due to their price advantage. However, this will most likely not be the case due to cost components for a completely installed system which are area related, for example mounting structures, cabling and installation cost. Especially with green field utility-power applications the area-related cost is inversely proportional to the efficiency of the modules used. The total system price p(tot) can be summarized as p(tot) = p(module) + p(power) + p(area) with:

   p(module): module price ex-works as discussed in the previous sections, plus a margin for the installer
   p(power): power-related costs like DC-AC inverters and some approval procedures
   p(area): mounting structure, cabling, cost of land and installation costs

In 2012 the majority of c-Si modules showed an efficiency of about 16% (based on total module area), while TF modules had an efficiency of about 10-13%. Total prices in Germany had been in the range of about $1.8/W for small/medium-sized systems mainly with c-Si modules. The end customer prices for the three components were approximately p(mod)=$1/W, p(power)= $0.3/W and p(area)=$0.5/W. If one keeps the p(power) constant for a given system size and also the p(tot), one can calculate the needed module price as function of the module efficiency, since the p(area) is inversely proportional to the efficiency. This relationship is shown schematically in Figure 7.8 on a relative scale.

From a simple calculation it can easily be seen that a module of low efficiency will need a lower price than a module with higher efficiency. In particular, for a 12% efficient Thin-Film module one would need ~$0.2/W lower module price compared to a 16% c-Si module for the same total system price. This is comparable to the price differences as seen for the various future PEC-scenarios in Thin-Film and c-Si module prices.

In power-related applications it is important to take into account which additional parameters are influencing the generation cost of a kWh. One

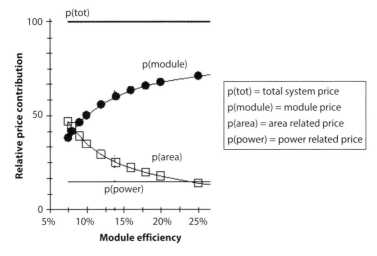

Figure 7.8 Dependencies of price contributors for installed PV systems on module efficiencies.

such parameter which has to be added to this comparison is the different temperature coefficients for efficiency. Today this is highest for c-Si modules and lowest for a-Si TF modules. It is also important to mention the fact that for applications where area related costs like mounting structure and installation do not have to be taken into account, like for façade installations (BIPV or Building Integrated PV) where the insulation glass window will be installed anyway or in a new roof, where the modules are replacing the tiles. As already discussed in the market chapter there is no "one size fits all" solution. Even if due to the significant overcapacity for c-Si modules the prices are artificially low and therefore Thin-Film products are hindered from increasing their market share I am still convinced that these products will penetrate the overall market.

7.5   Price Experience Curve for DC/AC inverters

Inverters are the heart and brain of grid connected PV systems. Their simplest task is the transformation of the PV module's DC into grid conform AC current. Knowing that the I-V- curve from the modules can change rather quickly depending on solar radiation and other climatic conditions, so does the maximum voltage and current of the modules. There has to be a fast and real time adoption which is called maximum power tracking. I remember when electrical engineers developed the first inverters in the late 1980s and they did so as they had been trained, namely to build a machine with the highest efficiency at nominal power. This was taken as the DC peak power of a given system in laboratory conditions, e.g. at 25°C and 1,000W/m². Not only is the temperature of the modules in the field considerably higher at highest irradiation compared to the laboratory conditions, but the histogram of different irradiation levels shows that this high irradiation is only observed a few hours per year and throughout the year most of the annual solar energy is obtained at much lower irradiations. As a result, the originally developed inverters had a rather low annual efficiency. This changed quickly when the real framework conditions were introduced to further development and today the annual efficiency according to a normalized histogram is approaching astonishingly high values of 98%. In addition, today's inverters must also be able to deliver more and more grid services[5]. It is not widely known that state of the art inverters contain the calculating power of a modern laptop. Although the technology became more and more complex the specific prices of inverters showed an impressive price experience curve as shown in Figure 7.9

For two power classes of inverters, typical decentralized string inverters below 20 kW and large central inverters well above 20 kW (typically 100kW), a PEC is shown with a simplified assumption that the prices for

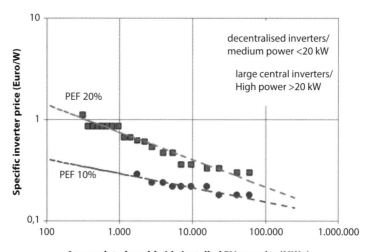

Source: SMA, Oliver Fuehrer 2011

Figure 7.9 Price Experience Curve for large central and decentralized medium power DC-AC inverters.

---

[5] Reactive power and fault-ride through.

the two inverter types were plotted for the respective total market volume (because it is difficult to correlate the price of the two inverter types with their respective cumulative volume). This implies that the "real" PEC would have a steeper slope for both curves. The important lesson to be drawn from the graph is that for small inverters a similar PEF of about 20% is obtained – in reality even slightly higher – as the one observed for solar modules. The bigger centralized systems which are still less expensive per power show a considerably smaller PEF of about 10%. It remains to be seen whether in the future we will see a crossing of the two extrapolated curves at around 1,000 GW with smaller systems becoming less expensive than bigger ones.

7.6   Price Experience Curve for Wind Energy and Other Relevant Products for a 100% Renewable World

When analyzing the specific prices for *on-shore wind mills*, there is much less decrease as function of cumulative sold wind power compared to the mass produced products described in the preceding sections. Starting at ~$4/W$_{wind}$ in 1980 there was approximately a $1/W reduction every 10 years to ~$1/W in 2010 (for on-shore only). Specific prices after 2010 even increased slightly, therefore the PEF in a wind-PEC even has a positive value now. This is no surprise since the larger windmills have higher pylons, longer blades and heavier gearing mechanisms which all add to the specific cost. Fortunately there is still a decrease in the LCOE for larger windmills as the output per installed power increases significantly as described in Chapter 4.2:

- higher pylon → higher average wind speed → power increase by third power of wind speed
- longer blade → power output increases with the square of blade length.

This is fundamentally different to the case of PV systems where on a given installation site the output of the PV system can only be increased linearly through efficiency, yet the decrease in LCOE declines according to the PEC of the various components (modules and inverters).

Future battery and fuel-cell products will also have a good potential for a 20+/-% PEF and a negative slope of the PEC, similar to PV modules. The main argument for this is the modularity and mass production of the individual cells assembled in the stack. Besides material considerations there is solid evidence that the levelized cost of storage (€ct/kWh stored electricity in batteries) and other new technologies for electricity production (e.g. €ct/kWh in fuel cells) will decrease similarly as in the case of PV.

For most of the other renewable technologies such as *hydro, geothermal and biomass* there are less mass produced units, which is the reason why they will not show a similar price decrease with increased cumulative volumes installed.

8

Future Technology Development

8.1 General Remarks on Future Technology Developments

Having seen the power of Price Experience Curves it is interesting, based on the state of the art of technology as of today, to look at how the different technologies will most probably develop in the coming years. Whenever there is the possibility for scaling up, adding new processes, switch to new and more cost efficient materials, increasing the efficiency and more, there is a high chance that prices will follow the respective Price Experience Curve with increased cumulative volume; and do not forget what was said in the past about the further development of well-known products – like semiconductors, flat panel displays, architectural glass and many more – which proved to be much too pessimistic.

8.2   Photovoltaics

8.2.1   PV Product Portfolio

In the late 1990s I proposed a potential development for the various PV technologies in terms of module price development versus module efficiency which is shown in Figure 8.1. It was already described that for energy production purposes the lower efficiency modules must have a lower price compared to the ones with higher efficiency in order to compensate the higher BOS and installation cost. The range in relative prices is shown for the different technologies for 2010 in the upper band, starting with Thin-Film technologies, then crystalline Silicon modules and on the right the high efficiency concentrating systems.

As time goes by the different module types will decrease in price – given by their respective Price Experience Curve – and increase in efficiency. Additionally, new technologies will emerge in addition to concentrating systems, as seen with the green dots representing dye and organic solar cells on the lower end of efficiency and price. It should be emphasized that this figure only summarizes a general development and should not be misunderstood as a quantitative forecast for price and efficiency development.

In summary I could envisage the following technological development:

❖ **Crystalline Silicon**
   According to the most recent ITRPV report [7-7] the module price has decreased substantially just between 2010 and

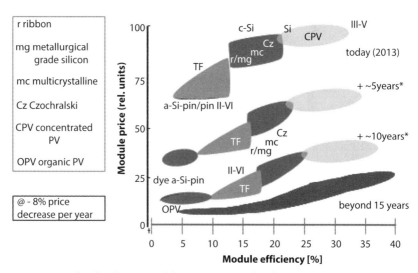

Figure 8.1 Further development of the various PV technologies.

2013 – mainly due to overcapacity within the whole value chain – towards $0.7/W at poly silicon prices of around $20/kg and a module efficiency for standard products of around 17%. Two of the bigger c-Si wafer, cell and module producers (JA Solar and Rene Sola) published their annual results for 2012 in March 2013. At a sold volume of 1.7 and 2.2 GW the turnover was $1.1 bn and $1 bn with a stated net loss of $276m and $203m for the two companies, respectively. If we postulate that for a healthy industry the profit should have been in the range of ~5+% we only achieve healthy module prices if the price paid for them had been 25 to 30% higher, which amounts to an increase of ~$ct15/W. Taking into account that most of these numbers do not reflect the full depreciation, the fair price of crystalline modules at last year's cumulative volume would be – as expected from the PEC – around $1/W. Most probably the global market volume would have been somewhat smaller with such prices but certainly not by that much.

With new cell architectures and novel processes and materials – both for cell and module manufacturing – module efficiency may well develop towards 25% in the late 2020s and reach a price level well below 50$ct/W. Not only efficiency improvements but also new and less expensive materials (e.g. replacement of silver contacts with Aluminum or copper) will contribute to the further continuation of the well-known PEC for c-Si modules. Let us make a rough estimate to see where this PEC may level off. At the end it is the .material cost (OPEX) which will dominate the cost structure. If we assume the following boundaries:

- Wafer size 200 × 200 mm² (back contacted cells), 80µm thick, 70µm kerf loss → ~14 g Si/wafer at $12/kg Si and 25% cell efficiency → **$ct1.7/W** plus **$ct2/W** other wafer add-on material cost
- Back contacted cells, efficiency 25%, cell add-on ~ **$ct2/W**
- Double glass (each 1.4mm and $5/m² → **$ct4/W**) modules with AR coating and 5% efficiency increase after encapsulation, back glass with thick film metallization for interconnect matrix, edge sealing losses leveled by encapsulation gain → 250 W/m² module power (corresponding to 25% module efficiency), encapsulation and interconnect

matrix ~ **$ct3/W** plus module finish (back-rail, connectors, cable etc.) ~ **$ct3/W**

When adding these material costs together we arrive at ~**$ct16/W** (or $40/m^2). With a number for all other cost items (depreciation, utilities, labor etc.) which is assumed to add minimum 20 and maximum 30% to the OPEX of $ct16/W we arrive at a total cost of about $ct20/W. Well before this price level is reached the PEF will start to decrease – my guess is that this could occur at $ct40/W which would correspond to a cumulative volume of ~2 TW (see Figure 7.7). With a decreasing PEF we would then arrive at the $0.2/W price level anywhere after 30 TW cumulative volume of sold c-Si solar modules. At these price and efficiency levels the LCOE of PV systems will be well below $ct5/kWh in sunny regions.

Laboratory results [8-1] demonstrate the possibility of combining both worlds of crystalline and Thin-Film technologies in the future, offering the possibility to increase efficiency to well above 30% by introducing multi band gap devices. This may then be the prerequisite for a further continuation of the PEC mainly driven by the increase in efficiency at an assumed rather stable cost per area.

❖ **Thin-Film Technologies**
II-VI compound technologies (CTS and CI(G)S) will drive module efficiency levels towards 15%, in the longer run even up to 18% (the laboratory cell efficiency for the CI(G)S technology is around 20% today). TF Silicon for the power sector will only have a chance if today's bottlenecks can be overcome: absorption enhancement for thin amorphous layers leading to a substantial decrease of the Staebler Wronski effect (higher efficiency) and improved productivity for layer deposition (decrease of depreciation cost). Multi band gap structures could help to boost efficiency. The anticipated price levels at which the PEF is likely to start to decrease and the PEC eventually to level off may be slightly lower compared to c-Si and are appraised to be $ct30/W and ~$ct15/W, respectively. A similar estimation for the corresponding cumulative volumes of Thin-Film modules (see Figure 7.7) would lead to ~ 200 GW and >2 TW, respectively.

As already mentioned in the preceding bullet point for the future of c-Si technology, here we may also see a combination of crystalline and Thin-Film silicon (and other materials) based technologies.

❖ **CPV (Concentrated PV)**
Commercially available devices have concentration ratios of between 300 and 500, even 1,000 has been demonstrated. It is therefore possible to further increase the efficiency of solar cells above the AM1.5 standard level. Solar cells with III-V compound materials have today demonstrated an efficiency of 44% in the laboratory and in production average cell efficiencies are well around 40% (both at ~300x concentration). With the same concentration this leads to module efficiencies of 32%. As discussed earlier, the introduction of even more band gap tailored cells opens the way to even higher efficiencies in future years which may well be beyond 50% for cell and 40% for module efficiency. This high efficiency is a great lever to decrease specific BOS cost and thereby to reach lower LCOE which will become less than 5$ct/kWh in those regions with high direct irradiation.

❖ **OPV (Organic PV) and DSC (Dye Solar Cells)**
The development of all organic solar cell products is strongly supported by the chemical industry. With the positive results already achieved we can expect a variety of new and additional customer needs to be served with these new products. This could range from the integration of solar cells into textiles, to housing and covers for many consumer goods and much more. Most probably OPV will not replace the before mentioned technologies when it comes to service life times of 30 years and more. The same is true for DSCs which have the specific USP (unique selling point) of providing true colors with these products. From my point of view there is no need to develop such cells with highest efficiencies – which is done effectively with the already mentioned technologies – but rather to make best use of the compelling color advantage.

Combining the future development one could speculate in the long run that the following situation would arise, as sketched in the lower band in Figure 8.1:

- the low efficiency products (efficiency ~(5 – 10)%) will be dominated by organic and dye solar cells by flexible roll-to-roll processes and will effectively serve the consumer sector, both indoor and outdoor,
- the medium efficiency products (~10 – 30%) will be Thin-Film, crystalline silicon and a combination of the two for the 100s of millions of decentralized BIPV/BAPV applications, many medium sized power stations in areas without direct solar irradiation and for mini grid power supply especially in countries lacking electricity infrastructure as well as many other solutions to provide electricity as well as
- the high end efficiency products (~30 – 40+%) which are dominated by highly concentrating PV systems and will best serve very large PV plants all around the world in places with high and direct irradiation.

The relative split between crystalline silicon and Thin-Film technologies in the past started with 100% crystalline silicon in the 1970s and early 1980s. Thin-Film technologies entered the market for consumer applications which resulted in a share of ~28% in the late 1980s and early 1990s. The tremendous capacity build in Asia starting in 2005 which was done almost exclusively with crystalline silicon, shifted the relative share back to crystalline silicon. With the inherent advantages of Thin-Film products (e.g. integrated series connection, large area deposition technologies, least material consumption per area) it is my firm belief that in future years the relative share of Thin-Films will again increase. It is worth speculating on the potential timely development regarding the relative share for the various mentioned technologies including "New Technologies" (CPV, OPV and DSC) as schematically shown in Figure 8.2. The illustration shown was based on my judgment around 2000 when I extrapolated the past development towards 2010 and beyond. The simplified assumption at that time was: (a) Crystalline silicon technologies will decrease their relative market share by 10% every 5 years, (b) Thin-Film technologies will increase their market share and (c) new technologies like dye and organic on the low and concentrating systems on the high efficiency side will gradually increase their relative market share as a function of time. As seen for 2025 – maybe even earlier – and beyond the clear cut differentiation between crystalline and Thin-Film silicon may disappear as indicated by TFX (Thin-Film like crystalline silicon technologies) or as described before there will be combinations of various Thin-Film technologies on top of c-Si wafers.

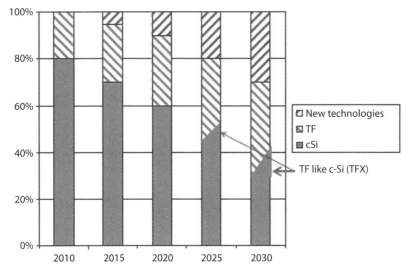

Figure 8.2 Technology split of the various technologies as a function of time.

The basic message of Figure 8.2 is not a timely and quantitative progno-sis for the relative share of the different technologies but the aim is rather to illustrate the earlier finding that depending on specific market develop-ments we most presumably will see all different technologies even in 20 years from now. This is why I do not support the idea of first, second and third generation technologies as this naturally would imply that while the third generation is developing the first one is disappearing.

In the long run after 2040 I could imagine a share of about (10 – 20)% low efficiency and/or shorter lifetime consumer products (OPV, Dye), 60+/-% medium efficiency power solutions (cSi, TF and a combination of the two) and ~(30 – 20)% high efficiency products (CPV III-V). The future split in technology and application also gives the industry the possibility to differentiate. This was a challenge around 2010 where c-Si standard mod-ules with only little differentiation served a market which was dominated by the financially driven FiT programs.

8.2.2   System Price Development

It is useful to have a simplified model available to obtain an approxima-tion of future system prices. We start from around 2010 and use the split according to the Marimekko-plot in Chapter 6.7, which for medium sized roof-top c-Si systems is ~60/8/12/20% for modules/inverters/BOS/instal-lation, respectively. With a system price of ~$2.5/W the $-numbers can be calculated. So how do we arrive at system price levels in 2020? For the module and inverter prices we follow the respective PEC curve and assume a growth number to obtain a cumulative volume. For modules we obtained in Chapter 7.4 a number of ~$0.6/W and ~$0.1/W as a "fair" price for modules and inverters, respectively. For the cost components BOS and installation there is no PEC. However, we can use the increase in efficiency as a factor which decreases the area proportional BOS and installation cost. For example if we take as an average in 2010 a module efficiency of 15% which may increase to 20% by 2020, this factor would be ¾. Productivity increases and more intelligent BOS materials could compensate for the increased material cost in the timeframe considered. By adding the four new numbers we arrive at ~$1.3/W system price in 2020. A similar exercise for the year 2030 gives a system price of ~$0.9/W assuming module efficiencies of 25%. This would result in a LCOE for PV electricity well below $ct10/kWh in Northern countries like Germany and well below $ct5/kWh in southern regions like Spain and Turkey. The above given numbers are summarized in Table 8.1.

For future new houses, where PV will be an integral part of the roof, these numbers decrease significantly because most of the BOS and installation is already part of the building. For large ground mounted systems the numbers given above will further decrease by 20-30%.

8.3   Wind Energy

In order to benefit further from increased power output with increased rotor blade lengths and pylon heights there will be a continuous development from today's existing 10 MW systems towards 15 MW, may be even 20 MW especially for on-shore applications.

Table 8.1 Future PV system price development and split for the major value added steps (small and medium sized c-Si PV systems assumed).

year	module efficiency	prices decrease according to PEC/PEF		price decrease by efficiency factor	sum
	[%]	Module [$/W]	Inverter [$/W]	BOS & installation [$/W]	[$/W]
2010	15	1.5	0.2	0.8	2.5
2020	20	0.6	0.1	0.6	1.3
2030	25	0.35	0.05	0.5	0.9

The challenges for off-shore systems will be long term stable operation under harsh conditions and sufficient auxiliary systems (e.g. special ships) to efficiently build the required numbers of such systems. As the mean wind velocity at sea is reached at much lower heights above sea level compared to on-shore systems there is no demand for such high pylons. The rotor size will then be an optimization between increased power for larger rotors and long term stability to minimize maintenance and repair.

A good summary of future cost developments in wind energy can be found by NREL [8-2]. In short, the authors project the LCOE from $ct(7–9)/kWh around 2010 to $ct5.5-7 in 2020 and $ct(3.5-4.5)/kWh in 2030.

8.4   Solar Thermal

As already discussed earlier less expensive materials will be developed for low temperature solar thermal systems in order to further decrease the cost per m² for the collector. Optimization at system level will additionally help to decrease the levelized cost of heat kWh. As new houses in the future will be well insulated and thereby will no longer need thermal collectors for heating and warm water, applications may shift to solar cooling and the provision of medium process heat for SMEs especially in southern regions.

The challenges for concentrated solar thermal systems have already been discussed in Chapter 4.3.3. With state of the art parabolic trough technologies the LCOE is substantially higher compared to PV and CPV. Whether Sterling dish or tower systems will be able to compete will be seen in the coming years based on the pilot projects being carried out today.

8.5   Other Renewables

There will be steady progress in all technologies – marine (tidal and wave), geothermal even hydro – but progress will not be as pronounced as described for PV and wind. It will be important for countries to make the best and optimal use of their resources.

8.6   Other System Components

8.6.1   Storage

Talking about electricity storage it must be emphasized that there are many different areas and technologies to be covered.

In *off-grid systems* storage is a must. Their size starts with so-called pico systems *(Wh)* to power the basic electricity needs of a household for the billions of families in developing countries. Combined with a small (5-10) W PV module a few LED lamps, radio and mobile phones can be effectively powered. Solar home systems with a 20-100 W PV module may go up to the *kWh*-range to allow adding refrigerators and computers. At a later time a multitude of solar home systems may combine to mini grid solutions with storage needs in the *MWh* range to also power small businesses and handicraft. The integration of small wind mills may complement the PV systems for this type of systems. A similar configuration can also be used for powering multi-MW power stations for remote industries like mining which today is done – more costly – with Diesel systems.

A similar range of different storage sizes will be needed for *on-grid systems*: from individual homes (kWh) to small industries/stores/hotels (MWh) to municipality level (GWh) up to seasonal/supraregional storage (TWh).

Especially the *kWh level* for decentralized storage will initially be pushed by the development of the electric car, although optimization is simplified for the stationary storage systems. The automotive batteries have to fulfill two conditions, namely getting cheaper per stored kWh and getting lighter per kWh. This does not have to be the case in houses, but the first condition is even more demanding, which could be achieved with a high weight optimized lead acid or other batteries. A recent report by Sandia [8-3] describes the development of new electrodes for lead acid batteries based on graphite materials which could boost the full discharge cycles from 1,500 up to 20,000 while only adding a minor cost to the kWh price of such batteries. It would be even more important to simply triple the cycles but at the same time to halve the specific production cost. If this could be transferred to mass production the price for a stored kWh of electricity would decrease from 25 €ct today to ~4 €ct. This would provide a disposable PV produced kWh in a private home.

For the larger *MWh to GWh* storage range, besides high temperature batteries (like NaS) large volume new material reflow battery systems may also be developed. Just as a reminder: in the older days – still the case in my cellar – many homes had a 6,000 liter oil tank installed; the same volume and weight could easily be taken for future energy storage purposes. I am convinced that this development has not yet started as it is not yet recognized as big business. Whenever this will be the case we will see a new and quickly expanding industry. Compressed air and mechanical fly wheels will also add to the product portfolio for the upcoming needs for electricity storage.

Seasonal – or at least a few months – storage in the *TWh* range may be provided by newly developed concepts like "Power to Gas", where a surplus of renewable electricity is used to split water by hydrolysis and combine the hydrogen with CO_2 to form CH_4 which can be fed into the existing natural gas pipeline infrastructure in many countries. This "Power Gas" as I would like to name it, could be used directly in combustion engines for long distance traffic (cars, trucks, ships and planes) and also to produce electricity with fuel cells. In a number of countries pumped hydro is also a cost-effective and additional means of storing large amounts of electricity.

8.6.2 Transmission

High power and high voltage transportation can be done with either HV-AC (high voltage – alternating current) or HV-DC (high voltage – direct current). In both cases it is possible to have above ground or underground transmission lines.

There is a lot of discussion about the increase of transmission capacity. Especially large (off-shore) wind farms do need such enhancements on a bigger scale - at least today and before new ideas are developed as will be discussed later. Before an additional new transmission line needs to be built with all associated planning and approval procedures, there are technical possibilities [8-4] for HV-AC lines (both above and underground) to obtain such an increase like:

- *Reconductoring*: use of new larger single conductors, or new twin conductors in parallel.
- *Increase Operating Temperature*: an increased current flow will lead to an increase in temperature of the standard conductors. However, the standard materials used only allow a limited temperature increase due to accompanied sag of the conductors which may cause contact to underlying structures like trees etc. With the use of "high temperature low-sag conductors" it is possible to further increase the capacity of the line as those new materials do not sag as much as the standard ones. The only draw-back is their increased cost (1.2 to 6 times) compared to the standard materials.
- *Voltage Uprating*: this approach is a very attractive option to increase the line capacity as long as the "Right Of Way" does not require too many increases.
- *Dynamic Line Rating*: most existing lines are designed for a worst case scenario with respect to environmental conditions, i.e. hot day with full sun and low cooling wind. The maximum design power flow level is defined in such a way that a further increase would heat the cable in a way that would induce sag

to unacceptable levels in the worst case environmental scenario. However, during most of the time these extreme conditions do not exist at the same time: if it is cooler, windy or
rainy the temperature increase due to higher current through
the line can be offset by the environment. As an example, an
increase of 10°C in the ambient temperature gives an increase
of the line of ~10% and an increase of the wind speed from no
wind to 1 mps (mile per second) increase the rating by 40%.
By adding information technology and on-line weather data
it is possible to significantly increase the power capacity by
varying the current flow according to actual conditions.

It should be remembered that methods to increase the power capacity by
increased levels of current (I) (this is the second and fourth method described
above) have one important draw-back: as the losses are given by I^2R, an
increase of the current by a factor of 2 will result in a loss by a factor of 4.

With the advent of large thyristors the transformation of very high voltage direct current (HV-DC) up to +/-800 kV became possible. This allows
electricity transport at significantly lower losses compared to HV-AC.
Typical losses are in the range of (5-6)% per 1,000 km. It must be noted that
for short distances HV-DC lines are more expensive while for longer distances they become increasingly less expensive compared to HV-AC lines[6].

8.6.3   System Services

In this chapter we have so far discussed the further technology development of important components (hard-ware) to power the world in the
future by 100% renewables. In parallel we must also implement surrounding system services including not only IT (Information Technology) as we
know it today but also the anticipated future development of IT and legislative frameworks (soft-ware). It will not be helpful to have an optimized
component available, for example a well-functioning and cost effective
solar module or wind mill, without any means to integrate it into the existing (more difficult) or future (much easier, if adequately foreseen) houses,
villages, counties, states and countries in the required quantity.

How would I perceive a future new district to be planned and realized?
Firstly, the streets and houses should be oriented in a way so as to capture most
of the sunlight through southern oriented large windows; the roofs should

---

[6] The reason for this is the additional investment in equipment to produce HV-DC from
HV-AC which has to be distributed over the length of the transmission line. Typically the
trade-off between HV-DC and HV-AC is reached within a few 100km.

ideally be monopitch in order to have the largest south-facing area available (architects will have to rethink since a north facing monopitch roof is standard); they should also not be disrupted with chimneys, roof integrated windows or other pieces which would not allow to integrate the PV systems in the future prefabricated roof. They will have a battery in their cellar which is also interconnected to the electric vehicle(s) of the family in two dimensions: one is to optimize the two battery systems and the other one to utilize the range extender of the electric car to provide power to the house or neighborhood when needed. Emphasis should also be placed on the fact that whatever is done should look nice and be perfectly integrated. What we see nowadays on some roofs is simply ugly: different sizes of solar heat and PV modules arranged around chimneys or other appliances which is really not the way to move forward. Even for a non-integrated roof-top PV and solar thermal system it is possible to have a harmonized and well-looking ensemble: as can be seen from the insert picture on the cover page which shows my own PV and solar thermal (the 3 modules on the left) system installed in 1994 where with a moderate additional investment the two technologies can be nicely put together on one roof and the retroactively added modules look from the distance almost integrated within the roof. Future houses will have proper insulation, triple insulated windows with low-e coatings and will produce within a year (almost) all the energy what they need. These future houses will then be integrated into a smart grid low voltage network. Through existing communication tools it will be possible to have a well-functioning offering and usage of electricity in this region, which could be controlled and managed at the municipality level[7]. Solar PV and wind parks in the vicinity of the respective low/medium voltage network will also be integrated. Weather forecast for wind and solar, Demand Side Management and local storage could already provide a high level of self-sufficiency. It is obvious that the infrastructure for new districts could be provided for the future rather easily, the implementation could then be done without difficulties and less cost at a later time. For the existing stock of buildings, unfortunately, it will take more time to integrate all these optimized ideas.

8.7   Importance of the Renewable Energy Portfolio – in Particular Solar and Wind

There is an easy and simple calculation which was done by Prof. Brendel from the ISFH which he allowed me to use. It impressively shows that

---

[7] This will create interesting and new business models for municipal utilities and public services

one always has to make a compromise in order to optimize the wider picture. In Figure 8.3 the simplified load curve within a year for Germany is shown using a 15% higher average load in winter compared to summer. The average solar irradiation and wind velocity was recorded for Potsdam in Germany [8-5]. Assuming a similar profile for the whole of Germany, the energy production for solar systems in summer is 11 times higher than in winter, whereas wind systems inversely produce 3 times as much in winter as they do in summer – on a monthly basis this already gives a good supplement for these two sources of renewable electricity production. Together with the German situation of a 15% higher load in winter compared to summer we have with P_W and P_S the total power for wind and solar, respectively:

$$\text{In winter: } P_W + 0.1P_S = P_{total,\ winter} = 1.15\ P_{total,\ summer}$$
$$\text{In summer: } P_S + 0.33P_W = P_{total,\ summerl}$$

From this we obtain the ratio of $P_S\ /\ P_W = 0.6$ and as each power unit of wind produces about twice as much electricity over the year for onshore wind, we obtain the respective energy ratio of $E_S\ /\ E_W = 0.3$. To find the best match for powering the complete load curve only through wind and solar, for each TWh of solar electricity produced, we would have three TWh of wind electricity. In terms of power this would mean that for every GW of solar power, we would have about 1.5 GW of wind power. If using 60 GW PV systems, this would mean 90 GW wind power. This simple calculation also demonstrates that energy-wise we would only cover about one third of Germany's electricity consumption. If we installed enough solar and wind to cover the total energy needs we would have to store a substantial part of these produced TWh.

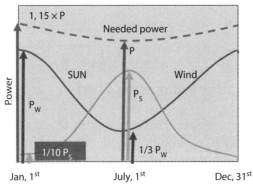

Figure 8.3 Simplified annual load curve in Germany and its supply by solar and wind.

Of course the energy and power needs in a country will not only be covered by wind and PV. It will be the optimized mixture of locally different renewable energy sources which have to be combined to a virtual power station. Only a few years ago this was done in Germany to demonstrate that 100% fluctuating state of the art renewable technologies could well serve the needs of power in Germany over the entire year. With technology advances this will develop very positively. The example shown is only meant to demonstrate in an easy and understandable way how a rough split of two renewable sources – wind and solar – in a particular region (here Germany) can contribute positively to the monthly load curve of this region. Today these simulations are done much more accurately by using load curves with numbers gathered in e.g. 15 minute intervals and also using wind and solar data with short temporal resolution.

9

Future Energy Projections – The 150 Peta-Watt-hour Challenge

9.1   Historical Development

It is interesting to observe the radical change with respect to the postulate that renewable Energies will be able to supply all global energy needs – and not in hundreds of years, but cost efficiently in the course of the current 21st century. When I started my industrial career in the early 1980s no one seriously dared make such a prognosis. Although I personally believed in a substantial – but at the time of the 1980s not a 100% – contribution of renewables I saw the relative role for PV as rather limited. In the 1990s with an annual market of ~50 MW I ventured a guess that PV may have up to 5,000 MW annual installations in 2010 thereby adding measurable TWh electricity – reality showed ~6,000 MW in 2008 (so my guess was obviously not too bad). The initial success of the Solar Thermal Electricity Generating Systems made people believe that this technology would be the large scale solar contributor while PV was seen as the junior partner. Wind started to demonstrate cost efficient production of renewable electricity. In the late 1990s I developed the global HVDC concept (to be discussed in the next chapter) to overcome the variability of wind and solar for a 100% renewable energy world. However, at that time I could only imagine the realization towards the end of this century. Cost efficient storage for electricity at the time was not seen as a viable solution. The cost and price development, in particular for PV, the high penetration into local grids already taking place today and the development of storage solutions (batteries, power to gas) created the atmosphere of a paradigm shift. Today it is en vogue for serious institutes like the Fraunhofer ISE and many others to present projections with 100% RE.

9.2   Some Future Projections and Scenarios by Others

9.2.1   Global Projections

It is no surprise that future energy projections in the 1970s/80s, that were carried out by traditional energy providers such as oil companies or global agencies like the IEA painted a picture with growing fossil and nuclear primary energy needs for the 21st century. Renewables at that time were only attributed a small contribution and the dominant share was expected to be provided by known hydro and biomass.

Almost every year a comprehensive data compilation, the *"World Energy Outlook (WEO)"*, is done at the International Energy Agency (IEA). It is a rich source for all kinds of energy related data from the past to the present. Based on past data they also make forecasts to the future decades. Until recently these projections did not assume renewables – beside hydro – to be important contributors to the future energy world.

In their 1994 World Energy Outlook the IEA stated that while "the further development of hydroelectric potential ... during the outlook period will be limited. ... Other renewable sources" (wind, PV, solar thermal etc.) "currently make a negligible contribution to the electric output mix. This is assumed to remain the case throughout the outlook period". The outlook period of this study was until 2010. Eight years later, the 2002 WEO gave some quantitative numbers up until 2030: the total Primary Energy needs were assumed to be 177 PWh with hydroelectricity producing 4 PWh, ALL other renewables (including biomass) 7 PWh and the lion's share was to come from fossil energy with 158 PWh (the remaining 8 PWh was nuclear). As a former executive director R. Priddle explained in this 2002 WEO [9-1]: "...new technologies" (=all renewables without hydro, the author) "will emerge on the energy scene within 30 years; but it will be much longer before they become dominant." This is also reflected in the numbers for the assumed world electricity additions between 2000 and 2030. The authors predicted ~2,700 GW new fossil power stations, 400 GW hydro and 100 GW nuclear, while they only estimated at that time 400 GW for all other renewables. Only 9 years later reality showed a drastic change in the renewable case: by the end of 2012 just wind has added 283 GW and PV more than 100 GW. Hence it did not take 28 years to reach the 400 GW but only 9 years. If we had included all other renewable technologies like bioenergy, solar thermal and geothermal, these 9 years would have even shrunk to a much shorter time.

It should be highlighted that within the last 10 years a significant change has occurred, not so much in the total Primary Energy predictions but in the predictions for the mix. In their WEO 2012 the IEA analyzed three different scenarios until 2035, which were called (i) Current policies, (ii) New policies and (iii) 450 Scenario. The relative contributions for fossil (coal plus oil plus gas), nuclear, hydro, bioenergy and other renewables changed noticeably and are equated with a future temperature increase of ~+5.3°C, 3.5°C and <2°C, respectively compared to pre-industrial levels. For an easy comparison these numbers are summarized in Table 9.1.

The decreasing share of fossil for the three scenarios and the increase in renewables (bioenergy plus all others excluding hydro) is clearly visible. Especially the increase from the WEO 2002, which stated 7 PWh (in 2030) for the renewable contribution compared to 40 PWh in the "450 scenario" (in 2035) is remarkable.

Nonetheless from my point of view these projections are still biased which is understandable when considering the history of this organization. Fortunately there has been a change in the WEO in recent years with

Table 9.1 Share of future world Primary Energy demand by IEA for different scenarios.

All numbers in PWh	WEO 2002	WEO 2012		
	2030	2035 current policies	2035 new policies	2035 450 Scenario
Fossil (coal, oil, gas)	158	173	151	108
Nuclear	8	12	13	18
Hydro	4	6	6	6
Bioenergy	–	20	22	26
Other renewables	7 incl. bioenergy	6	8	14
Total	177	217	200	174

regard to renewables but it is still – no surprise to me – overestimating the role of fossil, especially in view of the introduction of CCS (carbon dioxide capture and storage), which is deemed necessary for climate change reasons. I wonder what will happen when either this process does not keep its promises or simply becomes too expensive – certainly more expensive than the future cost of wind and solar.

The first report - "*The Limits to Growth*"- for the Club of Rome in 1972 was presented at the 3rd St. Gallen Symposium [1-1] and showed the world vividly that a global crash cannot be avoided in the 21st century if we continue in a business as usual modus with the growth of the global population, industrialization, environmental pollution, food production and the exploitation of natural resources. A number of scenarios were also discussed in order to prevent this catastrophe.

In 1998 E.U. von Weizsäcker, A.& H. Lovins [9-2] presented an important answer on how to proceed in a responsible way in their publication - "*Factor Four. Doubling Wealth – Halving Resource Use*"-. In the first part they gave 50 nicely described examples of how to achieve twice as much service compared with standard as of today systems with only half the resource input. In particular they highlighted the term "resource productivity" for this approach. Another 10 years later E.U. von Weizsäcker, K. Hargroves and M. Smith [9-3] highlighted an even increased potential and necessity for more productivity in their book - "*Factor 5. Transforming the Global Economy through 80% improvements in Resource Productivity*"-. This book impressively showed how to achieve a five-fold productivity increase which becomes possible with today's new technological developments. The examples for technology development described earlier in this book offer good support for these ideas.

In Germany an expert group, the "Scientific Council to the German government for global environmental changes (WBGU, Wissenschaftlicher Beirat der Bundesregierung Globale Umweltveränderungen)", was installed to advise the government on important global environmental changes. In 2003 a special report [9-4] was carried out which analyzed various scenarios for a future picture on energy with a barrier limit to keep the global mean temperature increase below 2°C. Most often a picture is drawn which shows ~310 PWh primary energy in 2050 (~450 PWh in 2100) with a share of ~50:50 for renewables:fossil in 2050 (75:25 in 2100). It should be mentioned that there are also other scenarios discussed with e.g. a primary energy need of ~200 PWh both for 2050 and 2100 and a share between renewables:fossil of ~50:50 in 2050 (80:20 in 2100) which would result in a secondary energy of about 165 PWh in 2050 (~185 PWh in 2100). In a recent publication from 2011 [2-7] they projected a 100 PWh renewable secondary energy model for 2050 which is shown in Figure 9.1. The last two scenarios (2003 and 2011) correspond to the first 4 bars in Figure 9.2. This picture shows two bars for each of the five different scenarios originating from different sources: the first one showing the split (renewable, nuclear and fossil) for the anticipated primary energy and the second (in blue) the resulting secondary energy for each scenario, respectively.

Greenpeace also started to develop a global energy scenario some years ago, entitled -*"energy [r]evolution – A Sustainable World Energy Outlook"-*,

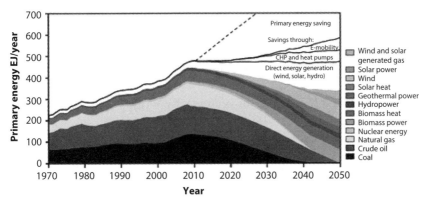

Figure 9.1 Historical and future development of Primary Energy (source: WBGU Flagship report, 2011 [2-7]; 500 EJ = 140 PWh).

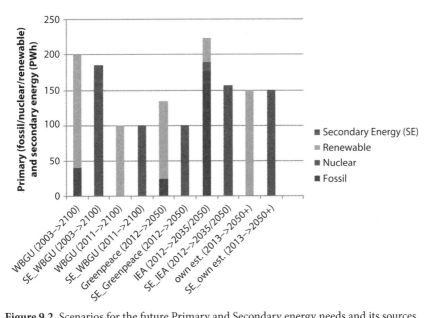

Figure 9.2 Scenarios for the future Primary and Secondary energy needs and its sources.

which is conducted in close cooperation with the EREC (European Renewable Energy Council). They regularly analyze the actual energy situation and forecast the energy sectors and sources for primary and secondary energy towards 2050. Interestingly they compare their projections to a so-called reference scenario which they derive from the respective latest IEA WEO data. As the latter ones in the recent studies only give projections towards 2030 or 2035 they use these data and continue with the trajectory towards 2050. In the last study [9-5] the numbers for the Greenpeace scenario for 2050 (bar 5 and 6) and the extrapolated IEA reference scenario for 2050 (bars 7 and 8) are shown. This "IEA reference scenario" calculated by Greenpeace is almost identical to the scenario from IEA in their WEO 2012 "current policies" for 2035 (Table 9.1). For comparison reasons my own model with 100% renewables (bars 9 and 10) is also shown (discussed in next section).

It is interesting to see that the secondary energy needs can be satisfied within a range of 100 to ~200 PWh. However, the split between the various contributions to the corresponding primary energy is enormous: while the IEA shows the major contribution from fossil and nuclear resources, WBGU (2011), Greenpeace and my projection show the clear opposite of the first one. It is only when, like in my and WBGU's 2011 proposal, all energy comes from renewables that we no longer need to look at the primary energy content.

In a fascinating book -"*2052: A Global Forecast for the next 40 Years*"- [9-6] the author J. Randers who also contributed 40 years earlier to the already mentioned report -"*The Limits to Growth*"- draws on his wealth of experience to give his most probable point of view for the coming 4 decades. Astonishingly – at least to me –he assumes a quickly reduced fertility rate which will lead, according to his model, to a maximum of only 8.1 billion people in the 2040s and thereafter even a decline of about 1% per year. I take this assumption to be a lower boundary of the global population. His forecast on energy needs and production are similar to the ones extrapolated to 2010 by Greenpeace based on IEA WEO data. This rather conservative energy picture can be understood by knowing how hard he was fighting for actions against climate change in the 1970s and the actual developments in 40 years – which are very frustrating. Here is the good news which I will try to sketch in my forecast: based on a much quicker price decrease due to technology development I am convinced that we can do it more quickly than the conservative projections suggest.

With respect to the future global population, which is an important parameter for the overall global energy needs – especially if we would like to allow everyone a similar level of energy consumption on average – I also consulted the UN numbers [9-7] where for the major regions in the world future projections have been stated as summarized in Table 9.2 for a medium fertility level[8]. From the information I analyzed so far I tried to estimate where a future global equilibrium could possibly evolve, which interestingly comes not too far away from the estimates by J. Randers – although some decades later.

9.2.2   Local Scenarios

In 1989 J. Nitsch and J. Luther published *"Energieversorgung der Zukunft (Energy Supply of the Future)"* [9-8] and analyzed the possible development for increasing shares of renewables in Germany from these early days until 2020. Although at this time there was no remarkable number of wind and PV installations besides hydro and biomass, their predictions matched the foreseeable contributions of all renewables in 2020 quite well.

Having discussed the global picture it helps a lot to see whether such a 100% renewables model could really work for an industrialized country like Germany. Photon developed a model -*"100% electricity by renewables in 2030"*- for Germany which was published in the October 2012 journal. Basic assumptions are 460 TWh end (electricity) energy (based on 2009 consumption and assumed to be the same for 2030, which could be attributed to energy savings for 2009 applications and these savings to be used for new applications like e-cars). The split in power for wind and solar was found to be optimal if ~240 GW wind (~530 TWh = 2.2 kWh/W_{wind} mainly

Table 9.2 Future global population (in millions) in major regions (2013, 2050 and 2100 data from [9-7], "equilibrium" case own estimates).

Region	2013	2050	2100	"equilibrium" ~2100+
Europe	742	709	639	~600
Northern America	355	446	513	~500
Latin America & Caribbean	617	782	736	~700
China	1,394	1,394	1,094	~800
India	1,252	1,620	1,547	~1,300
Rest of Asia & Oceania	1,691	2,207	2,141	~2,000
Africa	1,111	2,393	4,185	~3,500
Total	**7,162**	**9,551**	**10,855**	**~9,400**

[8] The UN report stated for the global population in 2100 a range between 6.8 and 16.6 billion people for a low and high fertility level, respectively

on-shore) and ~120 GW solar (~120 TWh = 1 kWh/W$_{PV}$) were installed (total energy 650 TWh). At 2:1, the ratio is slightly higher compared to 1.5:1 ratio from the aforementioned "Brendel" calculation (see chapter 5). This calculation only optimized power within a year in a very simple way, while Photon performed a 15 minute production and load simulation throughout the year. Additionally, the higher 2:1 ratio originates from minimizing storage, which was calculated at ~150 TWh electricity for P2G methane long term storage (plus little pumped hydro). In total the split would be 460 TWh electricity, 60 TWh generation management, 100 TWh losses due to PTG (useful heat), 30 TWh heat losses. One important lesson from this study is that it makes a fundamental difference in what way politics – together with the utilities companies – sets the target for our future electricity production mix. If we only allow 20% to still come from plants using traditional fossil primary energy we would not be forced to invest in storage technologies which are an essential part of the future energy system.

At Fraunhofer-ISE, H. Henning and colleagues developed a inter-sectorial (electricity, heat and mobility) modeling tool [9-9] with hourly time step simulation for a future energy system based on 100% renewables in Germany. Optimization was performed according to cost considerations. By means of energy efficiency a total of 1,450 TWh end energy consumption (compared to ~2,500 TWh today) was assumed. The split for the energy sectors was heating/hot water (~450 TWh), transport based on electricity (300 TWh, 30% electricity, 30% H$_2$ and 40% power gas from electricity), industry (~350 TWh) and non-energetic electricity (~350 TWh). The cost optimized production was elaborated to be 300 TWh PV (300 GW installed), 350 and 300 TWh wind on- and off-shore (200 and 85 GW installed) and solar thermal ~70 TWh (133 GW installed). The rest is biomass, useful heat from the Sabatier process (Power to Gas formation) and small hydro. Here we have a power ratio between wind and PV of only about 1:1 (energy ratio about 2:1).

9.3   Global Energy Scenarios and Market Development of the Major Renewables from the Author's Point of View

9.3.1   Simplified Projection for the Overall Picture

Bearing in mind that a quarter of 6 billion people (= 1.5 bn) consumed ¾ of the 140 PWh Primary energy we already calculated (¾ × 2/3 × 10 = 5) that a future primary energy of ~700 PWh would be needed, if we allow each of the 10 billion people in 2050+ the same energy use as we in our industrialized countries have today. These numbers are shown in the first line of Table 9.3. Such a high number was often used in the past to argue that so many PWh cannot be provided by renewables. However, these arguments simply forget two basic facts: firstly, the implementation of energy efficiency measures and secondly, the introduction of renewables where only the secondary energy in form of electricity or heat matters.

When looking at secondary energy we have 90 PWh today and without changing the pattern for the energy sectors we would then, under the same assumptions, need 450 PWh of SE in 2050+ (line 2). Assuming energy efficiency measures as discussed in chapter 3 are carried out, we would be able to reduce the SE needs conservatively to about one half, or 47 PWh, which would then only need 235 PWh ($45 \times 2/3 \times \frac{3}{4} \times 10$) in 2050+. The book by Weizsäcker et al. [9-4] which was already mentioned, stated that there could be an increased energy productivity of a factor of 5. This would give a lower boundary of only 18 PWh today and an astonishingly low 90 PWh in 2050+ for all 10 billion people. As seen in Figure 9.2 Greenpeace and also WBGU (2011) assume with 100 PWh secondary energy a similar number in their 2050 scenario.

After our two exercises in decreasing losses and reducing end energy needs for the "Better Quality of Life", we can now define our future secondary energy needs. The boundaries of required secondary energy for ~10 billion people who will all have a similar energy consumption are between 90 and 235 PWh. There are good reasons to choose for the future Secondary energy needs a number, which is about in the middle of the two boundaries. First, it may be too optimistic to install in all countries and all households

Table 9.3 Global energy needs as of today and in 2050+, assuming ~10 billion people and for various boundary conditions.

	Global need 2010 (@1/4 population needs ¾ PE) "inequity" [PWh]	Need in 2050+ ($\times 3/4 \times 2/3 \times 10$) "same energy for everyone" [PWh]
Primary Energy	140 as of today	700 business as usual
Secondary Energy (SE)	90 as of today	450 business as usual
SE with "mild" energy efficiency (factor 2 as in Chapter 3)	47	235
SE with "aggressive" energy efficiency: Factor 5 [9-4]	18	90

globally the concept of "aggressive energy efficiency measures". Secondly, if even recent IEA projections have Secondary energy numbers well below 200 PWh, it is too pessimistic to assume only the "mild energy efficiency" everywhere and for everyone. Therefore I suggest a future Secondary energy need of 150 PWh. The corresponding future picture for primary energy in form of the various renewables, secondary energy and end energy is sketched in Figure 9.3. I would be very surprised if the true number in the future were outside a +/-30% level (~100 – 200 PWh) of this figure (this of course assumes a peaceful world together with no serious natural disasters). We will see later that in the future we will have a much higher fraction of electricity than today. In the old world, where electricity needed about twice as much primary energy in the form of fossil and nuclear, such a development would have been a major problem. However, with renewables as our new form of primary energy we no longer have to worry about the losses associated with converting sun light or wind into electricity. These renewable Energies are – according to human time horizons – endless and need only to be converted effectively. It should be remembered from Chapter 2.1 that according to the conventional counting method for primary energy (second bar, primary energy DEEM) we have for the "Direct Energy Equivalent Method" the same number of 150 PWh for primary energy as for secondary energy. For the "Physical Energy Content Method" there would be a slightly higher number (~10%) compared to the 150 PWh

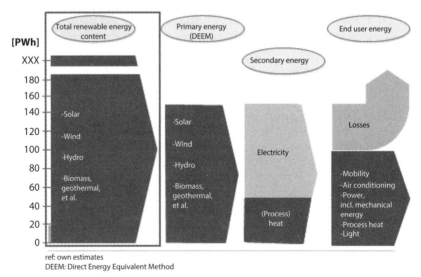

Figure 9.3 Annual primary, secondary and end energy in an energy efficient world with only renewables in 2050+.

depending on the proportion of geothermal and solar thermal electricity production. A rough estimate for the energy content contained in the renewable Energy Sources (indicated by "XXX" in the first bar surrounded by a red rectangle) can be obtained by assuming an average efficiency for all renewable energy converters (PV systems, wind mills etc.) in the range of 20%. In this case the 150 PWh of secondary energy would need 5 times as much or 750 PWh total energy content in the form of solar radiation, wind power etc. The fascinating fact is that these 750 PWh of energy per year are available for free and forever, according to human time scale. It should be remembered that the 150 PWh SE coming from renewables are only a small fraction of the sustainable potential for the annual renewable energy which was calculated earlier as ~3,100 PWh per year (the technical potential was even as high as ~80,000 PWh per year!).

There are still many experts who predict a rather small growth in electricity needs. This is understandable if only a small growth in renewables is considered. Even if only a growth from about 20 PWh electricity in 2010 to 60 PWh is assumed it would already imply a huge burden on the environment if these PWh were produced mainly by fossil power plants (nuclear plants are no longer considered because of cost, but also safety issues). The additional 40 PWh produced by fossil plants would need 100 – 120 PWh old primary fossil energy, about twice as much compared to the losses we have for electricity production today. Whether CCS will ever work is still unknown, but if so a clean coal kWh will be more expensive than renewable energy. The basic question of whether it will ever work comes from the consideration of how to store the CO_2 safely *for ever (at least for thousands of years)* as any leakage cannot be tolerated.

This picture changes completely if a high growth for renewables is assumed and the energy needs were powered by these renewables. In this case, we no longer have to worry about environmental concerns and thanks to the Price Experience Curve which will definitely also work for the various renewables there will even be a reduction of the energy bill compared to the situation today.

The main reason why I see electricity as becoming the dominant secondary energy provider originates from Table 9.4 which impressively shows that most of the renewable technologies produce this versatile secondary energy.

We therefore can expect that a major part of the required secondary energy needs will be powered by electricity – for simplicity reasons I assume (at least) ~2/3 of secondary energy to be provided by electricity in 2050+ as shown in Figure 9.4. The other 1/3 will be for (process) heat for industrial needs. Most if not all of the needs for non-e-mobility (planes, ships, trucks

Table 9.4 Renewable Technologies and Secondary Energy Produced (p&h: Power & Heat).

Technology	Secondary energy
Photovoltaics (mainly decentralized, partially centralized)	electricity
CPV/CSP (centralized)	electricity, p&h
Solar thermal (low to medium temperature)	(process)heat
Wind (on- and off-shore)	electricity
All other (hydro, geothermal, biomass, wave&tidal etc.)	electricity, p&h

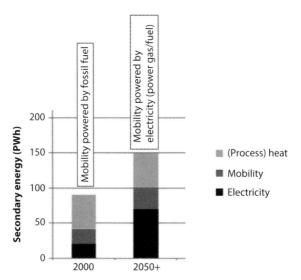

Figure 9.4 Timely development of the major energy sectors.

and long distance buses) will also shift to electricity as the source of secondary energy in the long run. This electricity will be used to split water into hydrogen (and oxygen) which could either be used directly in a fuel cell or reacted with CO_2 to form CH_4 ("power to gas" or short "power gas"), which could be transported like natural gas in our existing gas pipelines and infrastructure to be used in households and modern gas driven automobiles. Depending on the proportion of power gas versus electricity for transportation, we may have to install slightly higher capacities for renewable technologies to compensate for these losses. Just as an example: if we assume air traffic to be powered by power gas (CH_4 from hydrogen originating from electrolysis with electricity from renewables), we may need 2–3 PWh electricity in addition which according to our chosen example can be provided by adding slightly more renewable electricity generating capacity. This, however, does not change the overall picture for the future.

The relative contribution from solar radiation, wind and hydro plus all others is approximately 90%, 9% and 1%, respectively (for the technical potential this contribution is even 98.8%, 0.6% and 0.6%, respectively, see also Figure 9.5 and Chapter 4). From a simplistic point of view, one could assume the future energy delivery (150 PWh) – which is only ~5% of the sustainable renewable energy offering – to be proportional to this split. For future energy needs we have therefore enough resources available and can even pick those which will satisfy the required energy most conveniently and/or cost effectively.

This, however, would not take into account the local specific conditions for all other renewables besides solar and wind. This is the reason why in Figure 9.5 I boosted the 1% "all other" contribution to 20%, wind from 9% to 20% and decreased the 90% solar to only 60%. For my simple analysis I split the solar sector into three roughly equal parts: all the decentralized electricity contributions, best served by PV, the large electricity generating systems (CPV and CSP) in regions with high and direct irradiation and the solar thermal applications for heating and cooling purposes. Hence I arrive at a technology split which is easy to remember as 5 times ~20% each (see Figure 9.6). I do know that this approximation is by no means proof

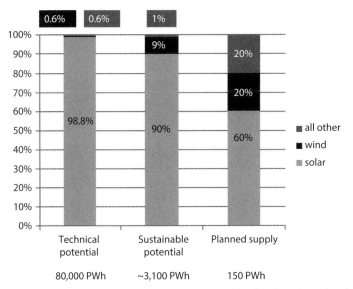

Figure 9.5 Split between solar, wind and all other renewables for the technical and sustainable potential as well as the assumed supply for SE in 2050+ (technical and sustainable potential [2-7], planned supply own estimates).

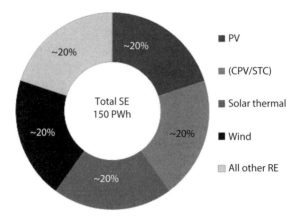

Figure 9.6 Split (%) of the various sustainable renewable energy resources for a 100% renewable world in 2050+.

for the simplified split – but it may serve as a top-down approach to understand what the future solar contribution versus wind and all others will be *at least* and why electricity will naturally be the most important contributor for a future renewably powered world. Most probably the contribution of PV is under- and the solar thermal is overestimated, but these differences are peanuts compared to the overarching picture. Let us not argue over which of the 5 areas mentioned is the biggest or the smallest but let us stand united and fight for the quickest schedule for the 100% renewably powered world.

9.3.2 Development of PV Market

In Figure 9.7 a very simplified projection for the future PV market growth is summarized. The first two lines show the average annual growth which took place during the two decades of 1990 – 2000 and 2000 – 2010. There is ample reason for the fact that the 50% annual growth in the past decade will not be replicated in future decades. The starting point was a 20 GW annual market in 2010. I assumed two different pathways for the coming 4 decades until 2050 which lead to a cumulative number of PV installations of 23 and 30 TW, respectively. With a conservative assumption of only 1.3 kWh/W_{pv} – which among others takes into account non ideal orientations of PV systems – this leads to an annual electricity production of ~30 and 38 PWh, respectively. The smaller number corresponds to the ~20% of the required total secondary energy in 2050+ for the scenario mentioned before and the higher number shows that a higher generation of PV solar electricity is also feasible to cover 20% of a global 200 PWh scenario mentioned earlier.

	V1	V2
Decade	% growth	p.a
1990–2000	20	20
2000–2010	50	50
2010–2020	20	20
2020–2030	14	15
2030–2040	8	10
2040–2050	4	5
Cumulative PV power in 2050 [TW]	23	30
Annually produced energy [PWh] in 2050 at 1.3 kWh/W (average)	30	38

Figure 9.7 Different growth rates for the global PV market and resulting cumulative market volume and annual energy production in 2050.

Combining the described future developments for the various module technologies and combining this with the otherwise needed components to complete a PV system, it is safe to say that in future it will always be cheaper to use existing substrates or structures and integrate a PV device and utilize the produced electricity compared to a not doing this. According to the two market growth models a volume of 600 GW per year may be expected as early as the mid-2030s. In order to gain a better understanding of what it takes to discuss an annual market of that size, the following examples show the different applications together with their respective volumes.

- **Consumer products (system size up to ~100 W)**
  Any housing of a device which uses electricity will have a nice looking module with a color according to the customer's wish. Outdoor textiles, rucksacks and tents will all have PV integrated to power the battery of a mobile phone or camera, keep the drinks cool and much more. Most cars will have a solar module integrated either into a cavity within the roof or as part of a sunroof. This will to a smaller extent to recharge the batteries of e-cars, but more so to drive a fan sucking cold air from below the car to push the warm air from the passenger cab to the outside to create a comfortable ambient air when sitting on a parking lot without shade in summer. For all

these consumer product applications I estimate 200 million systems in future with an average size 50 W each, which will result in *10 GW* per annum market. A huge market with 100s of millions of so-called "pico-systems" will develop for basic energy supply for the billions of people in developing countries. Due to energy efficient appliances, e.g. LED lamps instead of ordinary light bulbs, the system size is pretty small today. I remember when I was installing solar home systems in the 1980s we needed a module size between 20 and 50W; together with small batteries for 3 light bulbs, and transistor radio/black white television. The price of such a PV/battery system was in the range of $100 to $200. Today we can make use of energy efficient appliances and need only 5 to 10 W module size and improved batteries which decreases the price towards $10 to $20 – this is a tenfold decrease in price for the same quality of life! Together with micro-credit financing, which was also not available 30 years ago we can make a big change to the billions of people who otherwise would not have access to electricity. And let's not forget: communication with mobiles – unimaginable 30 years ago – is also possible today. Even if we consider 100 millions of such systems with an average size of 10W we only obtain an annual market of *1GW*. It should be highlighted that the impact per W_{PV} installed is most probably the highest for all those many people who otherwise would have no access to energy.

- **BIPV (building integrated PV)/BAPV (building added PV), (system size 100W to 100 kW)**
  PV modules will no longer be additions to existing houses but they will be building elements for the architects and the construction industry, be it façade elements, overhead glazing, tiles, part or all of the roof and, and, and... Of course there will still be standard hooked modules into balustrades and BAPV elements for existing houses. If we consider an annual global market of 25mio systems to be equipped with PV with an average size of 6kW each, we obtain *150 GW*.

- **Industrial applications (system size 100 kW to 10 MW)**
  For the millions of m²s of industrial buildings there will be efficient solutions to deliver as many kWh's by PV as

possible (3 million systems with 50 kW would give *150 GW* annually).

Another very big market is the off-grid solution for many remote industries and agriculture: in a first step adding PV to existing diesel generators can already decrease the average cost of electricity (PV-Diesel hybrids); depending on the load curve needed, most electricity can in future be supplied cheaper with PV plus storage (plus small wind generators) systems. Similar systems will also be used to provide village electrification in mini-grids in developing countries (this market segment could well even exceed *150 GW* annually).

- **Large green field PV systems for utility companies and municipalities (system size 10 MW to GW-range)**
  Today's largest PV system, the Agua Caliente project in Arizona (US), has an AC power of 247 MW at the end of 2012 and is projected to reach an installed capacity of 397 MW after its completion in 2014. There will be many more to come in future years and depending on the irradiation conditions adapted technologies will be used. The system size for these large PV systems will be in the several 100 MW -range and will reach the power output of today's fossil power stations (potential market for large field fixed tilt could also be above *150 GW*).

Adding up all market segments one can get an idea of what it means to discuss an annual market in the range of 600GW. In Figure 9.8 the split of the various market segments for a 600GW global annual market

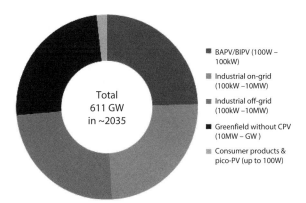

Figure 9.8 Market segmentation for an envisaged annual global 600GW market in ~2035.

is summarized. The cumulative volume of installed PV systems in the mid-2030s would be ~5,000 GW and global electricity production would be ~6,500 TWh.

Besides large hydro, PV and wind will then be the forms of electricity generation with the lowest LCOE figures among all renewables – and also lower than traditional fossil and nuclear technologies. Therefore there will still be further growth towards a 20% participation of the global secondary energy needed. The resulting annual market volumes would then be between 1,500 to ~2,000 GW in 2050+. For simplicity reasons I did not take into account the replacement of old PV systems. The main reason for this is that the lifetime for most quality systems will be 30+ years and therefore the number of installations in 2000 or 2010 with 0.3 or 30 GW, respectively, is fairly small compared to the annual production anticipated after 2040.

Another interesting topic is the regional development where all of the PV systems will be installed. Figure 9.9 sketches such a worldwide development of the respective cumulative installations and is simply based on the situation as it was (2010), reflecting the various regional support programs, towards a situation in 2050+, where the installations are proportional to the assumed population in the various regions (here taken from the "equilibrium case" in Table 9.2). While the column for 2020 is based on today's knowledge on short term market growth, the third 2050+ column is only indicative and describes a development based on

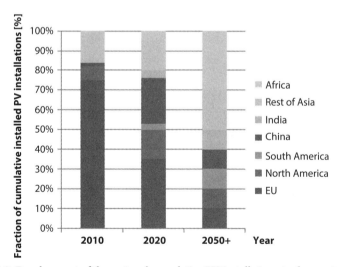

Figure 9.9 Development of the regional cumulative PV installations in the coming decades.

some major simplified assumptions: number of regional population, a similar energy usage per person globally and most of the regional energy consumption delivered by the same regional renewable energy supply. The challenge to reach the 30% relative contribution for the continent Africa is obvious.

Looking ahead towards 2050+ and analyzing the 150 PWh global scenario we would need the following numbers of PV systems cumulatively installed by then, as shown in Table 9.5. The range of system size for the different categories should be understood as described above for the market segmentation and is given in the table as a respective average.

Although at first sight the numbers may look large, they become reasonable once we understand that we are installing them all around the globe, in each and every country and within the coming 3+ decades.

9.3.3   Industrial and Company Policy Related Aspects for PV Industry

Combining the numbers for market growth (Figure 9.7) and price numbers (from Table 8.1) the turnover in the various industry sectors can be estimated as shown in Table 9.6.

Taking the further volume growth and considering the price decrease of the components as described in Chapter 8.2.1 we arrive at a total turnover of ~$800 bn in 2040 and ~$1,100 bn in 2050. These turnover numbers

Table 9.5 Number of PV systems needed for our 150 PWh reference scenario with 20% PV contribution (~23 TW x 1.3 kWh/W = ~30 PWh).

System category with average system size	Installed PV power [TW] in 2050+	Number of PV systems
ø10 MW green fields	~6	~600,000
ø1 MW industrial on-grid	~6	~6 million
ø1 MW industrial off-grid	~6	~6 million
ø10 kW BAPV/BIPV house systems	~6	~600 million

Table 9.6 Potential turnover in the PV sector until 2030.

	Market [GW/year]	Modules [bn $]	Inverters [bn $]	BoS [bn $]	Total [bn $]
2020	120	72	12	72	156
2030	450	158	23	225	406

will increase in these years if those for storage are included and may be compared to the ones in the automobile industry. In 2011 there was in Europe a turnover of ~$450 bn and the global one is about 3 to 4 times as big. Hence the turnover in the PV industry can become comparable to the global automobile industry in the coming decades. This should be for the various economic regions and their companies incentive enough to engage effectively in the participation of this important future industry. I do hope that also the industry in Europe will participate adequately along all added value steps – material, equipment and component manufacturing – together with the excellent R&D-institutes.

In many countries, particular in the Western hemisphere, there is today, unfortunately, only a short term thinking prevailing, driven by maximizing the profit and only focusing on one or few core activities. In some cases the addition of a new field is not followed even if substantial turnover at a lower profit margin is to be expected because there might be dilution of the promised higher profit margin with the standard business.

The other extreme can be seen with the attempt to create with a long time vision a highly integrated company. I have personally witnessed internally three examples. First the "Integrated technology company" by Edzard Reuter, CEO of Daimler Benz AG, who acquired in the mid-1980s the companies AEG, MBB, Dornier, MTU and Fokker to create a world-renowned technology company (when I created ASE in 1994 I benefitted from the combined solar activities both from AEG and MBB). This did not work as too many individual technologies were forced together. His successor Jürgen Schrempp tried an integration with the same industry – global car manufacturing – but the acquisition of Chrysler in 1998 and the integration with Mitsubishi in 2000 was not workable because too many cultures had to be respected (with the focus of only cars the 50% shares within ASE were sold to RWE). Lastly the "Multi Energy/Multi Utility" approach by Dietmar Kuhnt (CEO of RWE)) who added to the electricity business also an oil company (DEA), environmental activities (company for the disposal of waste), communication (Telliance), water business (Thames water) and an engineering group (TESSAG – it was this part of RWE, later renamed RWE Solutions, which was my shareholder until sold to SCHOTT). In all three cases – although for different reasons – the integration efforts detracted too much the focus from core activities and was not successful and led in each case to the destruction of many tens of $billion. A compromise between the two extremes to position a company in the medium term could help to take advantage of such developments as expected in the renewable industry.

9.3.4   CPV (Concentrated PV)/CSP (Concentrated Solar Power)

By 2050+ I expect large GW size centralized power stations in regions with high solar radiation to be making an important contribution to filling the 150 PWh energy needs. For regions with less direct solar radiation flat-plate PV systems will be the right choice while for those with high direct radiation the concentrated solar systems like CPV and CSP will be deployed. This is also based on the fact that in many regions we may have supranational HVDC super grids to transport electricity both from large off-shore wind farms and large CPV/CSP plants to the consumers starting in the 2020s and increasing thereafter. Alternatively to the supranational grids there may be a shift in energy consuming industries towards these large plants. If we assume an average size of 100 MW both for CPV and CSP, for simplicity reason each technology supplying half of the 20% = 30 PWh electricity and assuming 2,000 full load hours for those regions we arrive at a cumulative number of 75,000 systems each for both technologies.

- CPV will most probably grow after they will have successfully demonstrated their great potential for a truly cost effective electricity production
- CSP has principally the advantage of storing heat which can be utilized at a later time to also be converted into electricity in the turbines to better serve the load curve.

  - o Parabolic trough systems will only have a future if they can successfully replace the organic oil and introduce higher temperatures to increase efficiency.
  - o Solar tower systems have yet to prove maturity although they have by nature demonstrated highest temperatures with the potential for very high efficiencies
  - o Solar dish systems have successfully demonstrated their function although it is not yet clear which potential for LCOE they might be able to demonstrate.

9.3.5   Wind

- On-shore wind mills will still get higher and bigger to utilize the increase in efficiency (energy output per installed power) with higher wind speeds at higher pylons and larger diameters of the rotors

- Off-shore wind mills will not necessarily be as big as the aforementioned systems due to the harsh environment where they are installed. The most important factor will be the highest availability and lowest maintenance to make best use of the high number of full load hours in this environment. It may also be interesting to combine the rotor above sea level and a slow turning rotor driven by undersea currents below sea level at the same pylon.

In summary, in 2050+ we would have for our reference scenario 375,000 and 750,000 wind mills with an average power of 10 MW and 5 MW, respectively, for the on-shore and 760,000 wind mills with 5 MW off-shore. The associated assumptions are seen in Table 9.7.

9.3.6   Solar Thermal

- As discussed earlier I expect less need for hot water and heating in future decades because of better insulated houses, but there will certainly still be many million square meters worldwide for this application
- An important aspect will be solar thermal systems for cooling purposes although a similar argument applies for the future as described before
- Provision for process heat in the medium range for SMEs will be another area for future solar thermal applications

Although I may have overestimated the number of PWh to be delivered by solar thermal, it may be useful to make a quick check whether we could see a meaningful growth rate to arrive at a level of 30 PWh

Table 9.7 Number of wind mills needed for the main sectors to supply 20% of the 150 PWh reference scenario.

System category	Full load hours	Total wind power [TW]	Number of wind mills
On–shore 7.5 PWh with ø10 MW wind mills	2,000	3.75	375,000
On–shore 7.5 PWh with ø5 MW wind mills	2,000	3.75	750,000
Off–shore 15 PWh with ø5 MW wind mills	4,000	3.8	760,000

low temperature energy. Remembering the relationship between energy produced by installed power (700 kWh/kW$_{thermal}$) and that 1 million m² correspond to 0.8 GW$_{thermal}$, we would need 55,000 million m² to deliver the 30 PWh we need. With a cumulative installed base of 270 million m² by the end of 2010 we would need an annual growth of 14% until 2050 for the cumulative installations to reach those numbers. If there was a market to absorb these millions of m²s this could well be delivered by the industry.

9.3.7   Development of the Other Renewable Sources

We will not discuss in the same detail the further development for the other renewables but will simply highlight some ideas to be understood as food for thought. I will also not speculate – besides hydro with 8 PWh – how much each of the other renewables will contribute to the 30 - 8 = 22 PWh, which makes up the 5th pillar in our reference scenario of 150 PWh.

Hydro

- As already discussed, for many reasons there will be an asymptotic level of maximal capacity which was discussed in chapter 4.6.1 and will produce about 8 PWh

Bioenergy

- Not useful for electricity production on a large scale. Of course any 2nd generation biomass produced from waste should be used in the best possible way.

Geothermal

- For electricity production limited to only those countries with easy access to high temperature steam
- Well adapted in many regions for low temperature heat applications

Wave and tidal

- Like geothermal only useful in limited regions with good natural conditions

10

Likelihood of and Timeline for a World Powered by 100% Renewable Energy

10.1 Likelihood of a 100% Renewable World

It is my firm belief that if we only highlighted negative stories like global warming and finiteness of resources – as true as it is – in order to discuss a change from today's centralized fossil and nuclear energy supply, it would take quite long to get a global change organized. I am convinced, however, based on the superior cost development of all renewables compared to today's traditional energies that the financial argument, once understood by decision makers in politics, the financial institutions and industry, will make this change much quicker than most people would anticipate today.

The preceding chapters aimed to quantify the competitiveness of the most important renewable energy technologies in an energy efficient world. Besides the technology developments which most likely are the easiest to achieve there are some additional facts which are very often overlooked.

One important finding was developed by Stern in his famous report [10-1] which described the investments necessary for a renewably powered world compared to the damage which will occur due to climate changes if we continue to burn fossil and nuclear resources. He concluded that the damage was greater than the cost and the world would be in a worse shape by continuing with business as usual – melting of glaciers, rising sea levels with flooding in many highly populated regions, damage by stronger tornados, hurricanes and typhoons. In the latest report by the IPCC from September 2013 the increase of weather extremes and a quicker rise of the sea level compared to earlier reports strengthen this argument.

There has been and will be in the coming years a heavy debate on which of two very different concepts will be superior: When looking at past developments in most developed countries in the OECD world there has been a shift from an originally decentralized energy production in the 19th century towards centralized energy production, particularly for electricity generation. Big utility companies developed in many countries in the 20th century and dominated the energy infrastructure in these countries. Parallel to this development the "One-Way-Flow" of electricity from the big power stations via the various grid infrastructures to the consumers was established. We begin to realize as already discussed in a preceding chapter that the integration of renewables, in particular Photovoltaics with its decentralized nature, will need a much more decentralized infrastructure with a "multidirectional" flow of electricity to and from "prosumers"[9], still combined with larger power generating systems. It is easily understood that the larger the area is to produce electricity from renewables like wind and solar, the easier it will become to satisfy a respective load curve with less storage capacity. As it is not yet clear what generation and storage cost can be achieved in future years there is no clear cut answer to decide whether a Europe wide grid infrastructure together with large off-shore wind parks in the North and along the coast lines and green field PV parks in southern regions will have an economic advantage over a much more stringent decentralized production portfolio within areas with – for simplicity's sake – a radius of ~100km each. This would require considerably more storage, especially to cover the seasonal changes. The seasonal storage has the intrinsic disadvantage that one has to transform electricity into a storable energy form, for instance through electrolysis to produce hydrogen, which entails considerable losses. It makes a lot of sense to go ahead with the project P2G ("Power to Gas"), which uses electricity from renewables produced at times when the grid cannot absorb these kilo- to Terra-Wh and to

---

[9] This is the combination of what households in the future will be, namely producers and consumers = "prosumers" of electricity

transform them into methane via hydrogen plus CO_2 which can easily be fed into the existing natural gas pipeline infrastructure.

10.2   Global Network or Local Autonomy?

10.2.1   The Concept of a Worldwide Super Grid Versus the Hydrogen Economy

Moving towards a world which is powered at 100% by renewable sources needs additional features. Wind and solar are the most powerful and most cost effective ones but both have a strong seasonal and daily variation. The compliance with the load curve to be satisfied is an important aspect.

As early as in 1874 Jules Verne developed the vision of replacing the burning of coal with hydrogen and oxygen as a source of energy [10-2]. The term hydrogen economy was first used by John Bockris in 1970 and further developed together with Eduard Justi [10-3] in 1975 at a time when electricity transport was associated with high losses over large distances. Hydrogen is an elegant secondary energy which can be produced via electrolysis from water, stored over long periods of time and then be used either by combustion to produce power and/or heat or alternatively electricity through fuel cells. This is all correct and technically feasible and demonstrated but overlooks a simple fact: the losses associated with the hydrolysis and, if electricity is needed, the additional losses due to the conversion back by fuel cells. If we are not using laboratory numbers we may assume a loss of about 30% for both steps – this means that for one unit of electricity produced by renewables only half a unit of electricity remains after the twofold transformations ($0.7 \times 0.7 = 0.49$ or ~50%).

With the advent of HVDC, grid lines which operate very reliably today as subsea cables incur losses of only about 4 to 5%-points for every 1,000 km of transported electricity. As this technology will still improve in the future through technology development – higher voltages, improved cables and transformation processes – I foresaw a great future. The basic idea, summarized in Figure 10.1, which I developed in the early 1990s when I had to give a talk after a renowned advocate of the hydrogen economy is the following:

- If we start in a region that is open for a new way of energy supply and distribution, we may do this in Europe, enlarge into the Middle East and later into North Africa (EUMENA region). Such a grid would already span a time difference of about 4 hours, which would make it possible to use electricity produced in Eastern countries like Turkey in the early hours of the afternoon during the morning peak hours in countries like Great Britain. Of course all other renewables like wind from all areas would contribute effectively to satisfy the load curve. Such a development cannot be carried out in a few years but would realistically need a few decades. I was quite pleased when 20 years later the concept of an EUMENA Super Grid was taken up by the project DESERTEC. While I liked this part I was not at all satisfied with the project as such. The main reason for this is that when the project was announced in 2009 by the DESERTEC Foundation there was a lot of support from major utilities companies which liked this idea a lot as they could argue to be in favor of renewables, as they were using solar in more sunny areas like North Africa compared to Germany and would thereby stop the unwanted decentralized electricity producing systems in our own country – which take business away from them. In the meantime the Desertec Industry Initiative (DII) is focusing more of the production of renewable Electricity and its usage in the nearby region.

- The next region could be the NAFTA area (2) connecting the US and Mexico
- Several Asian regions (3) could then follow – or be developing in parallel to the NAFTA area
- Australia could follow (4)
- South America and Africa may complement the regional Super Grids (5)
- The last step could then be the connection (6) of the then existing regional Super Grids 1 to 5. This would allow the daily (East-West) as well as the seasonal (North-South) exchange of produced electricity with the regional load at a global level.

Many argue that such a worldwide grid is unrealistic because it would need too many countries to cooperate and because of its vulnerability. This argument, however, forgets what has been already achieved globally in the last decades with the development of gas pipelines and global oil tanker routes. If one plots these routes with a thickness proportional to the annually transported energy equivalent, there would be only a few very thick lines and everyone accepts the vulnerability of such concentrated

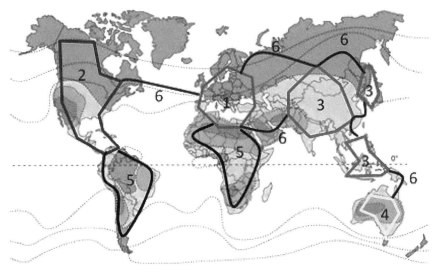

Figure 10.1 World Wide Super Grid as an alternative to a Hydrogen Society.

transports. The very positive element of a World Wide Super Grid shown in Figure 10.1 would be the multitude of grid lines – and not only the few ones shown – which would actually develop as a function of time. This would be much less vulnerable than today's situation. In addition it may be anticipated that as more and more countries realize the benefit of such a worldwide grid there would even be a desire to be part of it in order to gain access to a less costly energy supply for each and every one.

In addition, the losses associated with transporting either hydrogen or electricity are at least similar, most probably higher with hydrogen. Last but not least, the Worldwide Super Grid would not need large storage infrastructure, especially for seasonal storage. Although this worldwide Super Grid looks the most elegant there may be possibilities to even accelerate a 100% renewably powered world, as will be discussed in the next chapter.

10.2.2 New Horizons with Optimizing Regional 100% Renewable Energy Supply

It is interesting to observe how new technology developments and insights can influence and change one's own picture and perception. When I started my career in PV in the late 1970s, it was not foreseeable that wind and solar would develop as quickly as they did in reality. The concept of a (northern!) country like Germany to have a significant contribution of secondary energy delivered economically mostly by PV and wind as early as 2030 was simply out of reach. Should, however, the technology development which we experienced so far continue – which is more probable than not – we would be able to discuss completely new global concepts for energy supply more than ever before and this would make the global 100% supply by renewables even easier and more quickly doable.

By our own country experience in Germany we know how difficult and lengthy the process is of getting approval for a new HV transmission line. Approval procedures for enhancing the grid connection between Spain and France took about 20 years. It will doubtlessly be even more difficult – not technically, but due to approval procedures – when it comes to the interconnection of the various regional Super Grids shown in Figure 10.1 as indicated by the connecting lines (6). As a consequence, it may take several decades longer to introduce a global 100% renewable energy supply using this Worldwide Super Grid than it would take to individually build up 100% or only 90+% renewable energy supply in the regions 1 to 5 as shown in Figure 10.1.

If a more advanced sub region could demonstrate that a 100% renewable supply is superior both in terms of environmental as well as economic costs, then this could trigger a much faster spread for the 100% global renewable supply rather than a situation in which the regions have to wait until the interconnections are up and running. In the meantime they still have to use traditional fossil energy for their secondary energy supply.

10.2.3   Local Autonomy: Silver Bullet for the Decentralized Private and SME Sector Plus the Centralized Energy Intensive Industry

I would like to again give credit to Wolf von Fabeck, the mastermind for the support scheme "cost efficient feed-in tariff", who in recent times advocated for a strict local autonomy without additional transmission lines. The more I engage in the 100% RE scenario the more I appreciate the local autonomy – at least for the private and SME/office sector. Sometimes it may help to bring ideas together which today are still seen as irreconcilable antipodes. Today, the discussion is taking place in the traditional sector that the major energy contribution should come from wind off-shore which needs many additional transmission lines to connect to the centers of usage of energy far away from the costal lines in many countries – like Germany. Secondarily it is often argued that the energy intensive industry suffers disproportionately from the more expensive small and decentralized renewable energy production systems.

There is a simple loophole to escape from this dilemma, especially when thinking on a global scale. Like the aluminum production in Iceland where bauxite (aluminum ore) is transported with ships to a hydro plant, or the reduction of quartz to metallurgical silicon in Scandinavia which also takes place near a hydro power station, I could imagine a development in which energy intensive industries not only settle near to hydro power stations but also near to the coast where off-shore wind parks provide electricity or large solar thermal and PV plants in sun-rich areas deliver electricity and/or process heat, especially if such places can be easily reached by ship transport. At the end it will be the economic decision whether to transport electricity from large central power stations to energy intensive industries or to transport material to be processed to the power stations. Either way such a development would take away a lot of unnecessary discussions which we are having today. Such a relocation of industrial companies would fortunately not imply that countries would have to give up those industries but only that they would have to make use of the respective natural resources.

In contrast to this future "local autonomy for energy intensive industries at places of centralized renewable energy production plants". we will have in parallel the "local autonomy for the private sector and regional energy needs for offices and SMEs". The latter can well be equalized with the future smart grid areas that were discussed earlier.

10.3   Timeline for a 100% Renewable World

"Strive for Mission Impossible and be surprised"; with Mission Impossible being 100% global energy supply through renewables by 2050. I am realistic enough to know that most probably this "100%" will not be reached. However, if in selected regions we now start working towards this plan locally, we may be surprised to see that other regions may quickly follow once they realize the economic and environmental benefit in the first region. Yes, the front runner may have the burden of having spent more money on this conversion process for each power and energy unit compared to the followers – but it may also have the advantage of more quickly growing an industry to serve these energy markets with all the required products and services everywhere.

It may be helpful to also address the question of the timeline by looking at how in the past the global industry has developed (we already mentioned earlier such an analysis of the timely development of secondary energy forms by Marcetti). Three different timescales are discussed in this area: short economy cycles ranging from between 3 and 7 years ("Kitchin-cycles"), medium term economy cycles of up to 11 years ("Juglar-cycles")

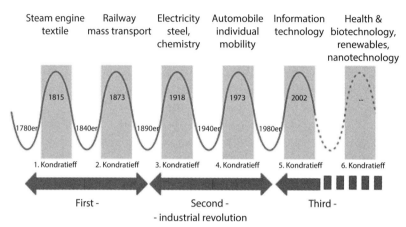

Figure 10.2 Technology cycles from the past to the future.

and the often discussed long time cycles of about 50 years, introduced by the Russian economist Kondratieff in the 1920s which he called "the theory of long waves". Joseph Schumpeter renamed this phenomenon "Kondratieff – cycles". In analogy this methodology was extrapolated by economists like L. Nefiodow [10-4]. The development and the major contributing technologies are schematically shown in Figure 10.2. Sometimes two of the 50 year cycles are also condensed into ~100 year cycles and highlighted as times for the 1st, 2nd and 3rd industrial revolution. The first phase is characterized by the basic innovation of the steam engine for mass production in the textile industry, and the introduction of the railway with the advent of mass transportation of people, materials and products. The second phase saw electricity, steel production and chemistry as basic innovations; with the automobile, driven by petrol, the development of individual transport. The third and still on-going phase started with the introduction of information technology, e.g. the internet, and the topic of "health for mankind" in a broader sense: not only in a narrow medical and constitutional sense but also psychologically, ecologically and socially. In this context environmental technologies like the introduction of renewables, energy efficiency, new transport technologies (like e-mobility), new materials and many more should also be named.

Extrapolating the duration of ~50 years for each Kondratieff cycle one could speculate that the running 6th cycle may peak in ~2050 and then develop into something new – the 7th Kondratieff cycle with radically new ideas on how the 10 billion people may live in harmony after the accomplishment of the 6th cycle: affordable energy from renewables without any destruction of nature and new materials made by nanotechnology without any limitation due to scarcity. The start of this 7th Kondratieff cycle could then be the start of a new and fourth industrial revolution with new materials and technologies we can only speculate about today. Whatever this future will bring, it could tell us at least that with the termination of the 6th cycle the energy question with respect to sustainability, ecology and economy will be solved by the second half of this century. It is my firm belief that this solution will no longer consist of traditional energy providing technologies based on fossil and nuclear, but will be based on the portfolio of renewable technologies.

Another important aspect for a quick change towards a 100% renewable energy world was elaborated by Nicholas Stern [10-1]. In his famous review he postulated that it will be far more expensive to continue with business as usual (BAU) compared to quick actions to stabilize the CO_2 concentration. In short, it was stated that only 1% of global GDP is needed annually to keep the temperature increase at around 2°C until 2100 compared to costs caused by damages to in a BAU model which would be 5 times higher with a potential of even being 20 times higher! Of course these numbers strongly depend on assumptions such as long term discount rates, considered time scales and much more. Some critical opponents like W. Cline [10-5] claim that some of the assumptions are too optimistic and overestimate the above stated advantages, yet they still come to the same basic conclusion that it costs more for BAU compared to investing in CO_2 stabilization technologies such as renewables. Of course, there are also people like J. Delingpole who may be an excellent writer and journalist and, according to "The Telegraph", the author of fantastically entertaining books. However, he unfortunately belongs to the kind of people who simply deny the findings of Nicholas Stern in a similar way as was described earlier in the case of the Climategate group which argues against global warming by man-made actions. To me this type of cynicism does not add to a decent discussion on such an important topic. Fortunately we live in a part of the globe where everyone can have their own opinions and can make them public without fear of punishment. But that freedom should not be misused to disqualify other opinions just by a mixture of half-true statements and discrediting other people.

More specifically I could foresee the following chronological development:

- By the latest in the 2020s, all of the described and necessary renewables such as decentralized PV, centralized CPV and CSP, wind on- and off-shore, solar thermal as well as all other renewables will have demonstrated their superiority over traditional fossil and nuclear technologies.

- Storage solutions which will both help e-mobility and fluctuating renewables to cover increasing shares of the energy needs of countries will also be available in this time frame.

- In some regions supranational HVDC transmission lines will gradually be added in the 2020s and 2030s to efficiently distribute the electricity from large wind off-shore and CPV/CSP plants. Alternatively we may see the relocation of energy intensive industries to be closer to the large centralized PV and wind power plants.

- The developments described above will not happen in any place or region around the globe now or in the coming years. The frontrunners will be Europe, the US, Japan and China and it will be then a question of time until more and more other regions will recognize the monetary advantages of introducing renewables and replacing traditional energy sources.

- Ideally the global energy could be served renewably at 100% in 2050. If energy efficiency measures could be accelerated so that only the lower boundary of 90 PWh is needed, this could already be achieved even earlier. However, as most often in the real world, existing systems – like traditional fossil and nuclear power plants – and traditional behavior will most probably and unfortunately prolong this time scale. But the bigger the gap between the lower cost of renewables and the high cost of traditional energies will become, the quicker it will happen.

11

Conclusion: The 100% Renewable Energy Puzzle

The most important arguments for a future 100% renewably powered world are summarized in Figure 11.1. There are basic objections to a "Business As Usual" scenario as summarized in the darker and reddish elements of the puzzle:

- Nuclear is a "no go" because of safety considerations which are not 100% guaranteed, and because of the unsolved waste storage problem. Nuclear also has no economic future: based on recent discussions between EDF (France) and British ministries the cost of newly built nuclear power station is such that an agreement was made to pay €ct11.2/kWh for 35 years for each produced kWh plus adjustment for inflation. In addition, uranium resources for fission reactors are quite limited and could only be stretched if reprocessing is considered, which would in turn cause additional safety and proliferation problems. Furthermore, the use of 3rd and 4th generation reactors like fast breeders would only add significantly to the safety problems. Fusion reactors are not being considered

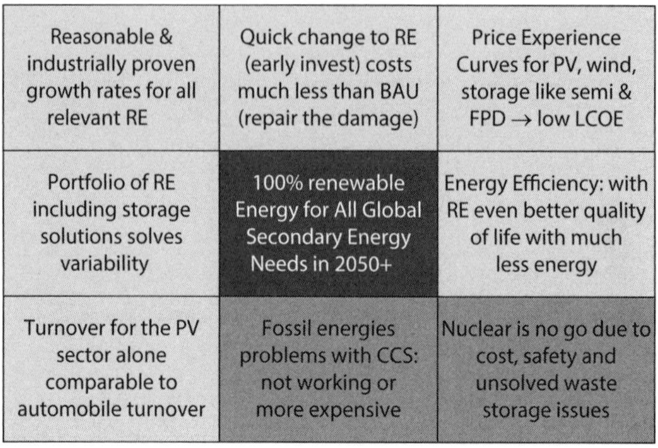

Reasonable & industrially proven growth rates for all relevant RE	Quick change to RE (early invest) costs much less than BAU (repair the damage)	Price Experience Curves for PV, wind, storage like semi & FPD → low LCOE
Portfolio of RE including storage solutions solves variability	100% renewable Energy for All Global Secondary Energy Needs in 2050+	Energy Efficiency: with RE even better quality of life with much less energy
Turnover for the PV sector alone comparable to automobile turnover	Fossil energies problems with CCS: not working or more expensive	Nuclear is no go due to cost, safety and unsolved waste storage issues

Figure 11.1 The 100% renewable Energy (RE) puzzle.

because even if technological challenges were to be solved, I am convinced that firstly they would not be feasible for the preferred decentralized energy production technologies and, secondly, they would be more expensive than renewables.

- Fossil energies pose either an environmental problem when CO_2 is released and contributes to global warming at unacceptable levels, or a cost problem if CCS is considered which adds significantly to the LCOE –if the long term storage can be solved at all. The topic of climate change is summarized well by Rahmstorf and Schellnhuber [2-12]. Traditional exhaustible energy sources have the natural disadvantage that despite new findings today they are definitely limited tomorrow and will bring with them increasing costs for unconventional future sources (e.g. deep water oil and gas reserves in the arctic sea). Once the time of peak gas and oil is passed, markets will react with increasing prices.

Fortunately there is good evidence that a future energy supply can be provided environmentally safe, economically superior, and with security of supply regionally, globally and forever. The major arguments for this are summarized in the greenish elements:

- Although not directly coupled with RE, energy efficiency measures for all appliances and products are important for a future world. This will help to have the same – with renewables even better – quality of life with less secondary energy – with renewables even eliminating the problem of energy losses through conversion from exhaustible primary to secondary energy. The ideas by Weizsäcker *et al.* "Factor 4: doubling prosperity – bisected consumption of nature" and "Factor 5: the formula for sustainable growth" summarize this topic impressively.

- The portfolio of all renewables – most importantly solar and wind due to their big technical, economic and sustainable potential – will already smoothen the energy delivery compared to the load curve required by consumers. The challenge of variability of the various renewables will be solved through the storage of electricity directly in batteries and pumped hydro as well as indirectly with power gas from surpluses in renewable power. Most likely a dramatic change will take place with regard to distributed and centralized renewable energy production and its respective usage. The decentralized electricity production through PV and local on-shore wind mills as well as heat by solar thermal systems will develop a situation of greater autonomy for private households, SMEs and offices in local areas. This will take place with the support of small and medium sized storage systems and the integration of the range extender from electric cars for power balancing. The large and energy intensive industry (metal smelters, chemistry etc.) will either concentrate close to the centralized power stations by hydro dams, off-shore wind farms and very large PV parks or, when economically more attractive, high voltage DC grids can transport electricity to the respective industry.

- Besides the fact that renewable primary energy sources (solar, wind, hydro, bioenergy etc) are abundantly available, the decrease of the LCOE for the various energy converters like solar modules, wind mills etc. is widely underestimated. This can be predicted on the basis of the Price Experience Curves for the various energy converters that contribute to the continuous further cost decrease similar to the well-known examples of semiconductor devices, flat panel display products and many more. Similar considerations hold true for battery systems and fuel cell devices. Table 11.1 summarizes the previously discussed LCOE's for PV and wind and estimates electricity storage for the time 2030+. The LCOE for clean coal and nuclear fission is also given for comparison. It is interesting to compare the sum of wind or PV plus storage with the traditional LCOE, which is roughly the same with one important difference: while there is no proven fact today whether CSS will really work and if so at

Table 11.1 Estimated LCOE's in 2030 + for various technologies.

Category	Technology	LCOE in today's currency [$ct/kWh]
Traditional	Clean coal with CSS	~10
	Nuclear fission	>~10
Photovoltaics	Southern areas (~2 kWh/W$_{pv}$)	3 – 4
	Northern areas (~1 kWh/W$_{pv}$)	6 – 8
Wind	On-shore (~2 kWh/W$_{wind}$)	3 – 4
	Off-shore (~4 kWh/W$_{wind}$)	4 – 5
Storage	Small (~kWh+)	6 – 8
	Large (~MWh)	<5

what additional cost, the renewable technologies are based on solid projection from technology development of the past with PECs looking towards the future.

- Although it is by no means scientific proof but rather a well observed phenomenon, the Kondratieff cycles are continuing, in particular the current 6th one which is mostly associated with "Man's Health" in a broader and more general sense. For example the introduction of RE is one of the important features for this current cycle which should come to an end around 2050 by analogy to the first 5 cycles. This would imply that by then the economic large scale introduction will have happened and the beginning 7th cycle will concentrate on new challenges. If we calculate average growth numbers for the various renewable Energies from today to the PWh necessary to provide 100% energy supply by 2050 (150 PWh) we can see from Table 11.2 that no unrealistic growth has to be anticipated – just remember the average growth for PV in the first decade of this century, which was well above 50% per year.

- Even if some RE technologies today show a slightly higher LCOE compared to traditional technologies, one must not forget that traditional energy production will become more expensive while RE will decrease their LCOE continuously (see Table 11.1). Additionally it is not highlighted that e.g. fossil energy receive more support money compared to the whole RE sector: the IEA WEO 2010 report states that in 2009 312 billion $ were spent globally for consumption subsidies

Table 11.2 Anticipated CAGR (Compound Average Growth Rate) for renewable Energies from today to 2050 (own estimates).

	~ 2010+/-		~ 2050+			
	GW	TWh	TWh	TWh	GW	CAGR [%] p.a.
Photovoltaics	~ 100	~ 120		~ 30,000	~ 23,000	+ 14.8
CPV/CSP	~ 1	~ 2		~ 30,000	~ 17,000	+ 27.2
Solar thermal	~ 190	~ 130		~ 30,000	~ 44,000	+ 14.5
Wind	~ 280	~ 600		~ 30,000	~ 10,000	+ 10.3
Bioenergy	–	~ 14,000	↓ ~10,000			substitute
Hydro	~850	~ 3,500	↑ ~ 8,000	~ 30,000	–	+ 2.1
Geothermal, wave & tidal, etc	–	~ 5	↑ ~ 12,000			+ 21.5
Total	1,421	18,357		150,000		+ 5.4

to fossil fuel while only 57 billion $ were given as support to ALL renewables. It is also mentioned that the internalization of external cost has been ignored. Alternatively, as described by N. Stern, each dollar of what is not invested in RE today has to be paid at a substantially higher level at a later date due to the damage caused by on-going traditional energy technologies (storms, droughts, sea level rise and many more).

- After all the renewable pathway opens also a new large scale industry in all areas. Just for the PV sector alone it was estimated that with the growth rates and price developments assumed the turnover becomes in the 2040s comparable to today's global turnover in the automobile industry, which is a $1.5 trillion industry.

- As a consequence of all surrounding elements we arrive 'almost automatically' at the central element in the dark green middle of the matrix, which calls for 100% renewable Energies for the global secondary energy needs in the future. The interesting question remains as to when this will have happened. Knowing that in Germany a number of municipalities have started to set a precise date around the 2020s to have a 100% RE supply, one could envisage that Germany could indeed demonstrate the same as a whole in the 2030s. It very much depends firstly on what those municipalities are able to achieve in terms of cost and, secondly, on how the still existing old and not optimized building stock can be changed efficiently. Should, however, an economy like Germany indeed be able to demonstrate that a change to 100% RE is not only positive for the environment and human health but that it is also in the foreseeable future – which can be easily calculated – superior to the incumbent traditional energy system, it would certainly trigger an international imitation. During which decade the 100% RE world is to happen will be subject to joint discussion – for all of us who are still around –in the decades after 2050.

References

[1-1]   D.H. Meadows, D.L. Meadows, J. Randers and W.W. Behrens III, *The Limits to Growth*, ISBN 0-87663-165-0 (1972)

[2-1]   IEA, *Energy Technology Perspectives* (2008), ISBN 92-64-04142-4

[2-2]   IEA (International Energy Outlook), World Energy Outlook (2011)

[2-3]   C. Marchetti and N. Nakicenovic, *The Dynamics of Energy Systems and the Logistic Substitution Model*, RR 79-13, International Institute for Applied Systems Analysis, Laxenburg, Austria, (December 1979)

[2-4]   C. Marchetti, *Society as a Learning System: Discovery, Invention and Innovation Cycles revisited*, RR-81-29, (November 1981) http://www.cesaremarchetti.org/archive/scan/MARCHETTI-032.pdf

[2-5]   Greenpeace, *World Energy [R]evolution: a sustainable World Energy Outlook* (2012), www.energyblueprint.info/fileadmin/media/documents/2012/ER2012_final_including_IRENA_foreword.pdf

[2-6]   EU FP6 Integrated Project QUANTIFY, *Quantifying the Climate Impact of Global and European Transport Systems*, (2010) http://www.pa.op.dlr.de/quantify/

[2-7]   Advisory Board to the German Government on Global Change (WBGU), Flagship Report 2011, *World in Transition: A social Contract for Sustainability*, ISBN 978-3-936 191-37-0 (2011) http://www.wbgu.de/fileadmin/templates/dateien/veroeffentlichungen/hauptgutachten/jg2011/wbgu_jg2011_en.pdf

[2-8]   David Carlson (2008), personal communication

[2-9]   IPPC (Intergovernmental Panel on Climate Change) Fourth Assessment Report, *Climate Change 2007*

[2-10] J.C. Comiso and F. Nishio, *Trends in the sea ice cover using enhanced and compatible AMSR-E, SSM/I and SMMR data*, Journal of Geophysical Research, 104, 20837-20856 (2008)

[2-11] J. Turner, J.C. Comiso, G.J. Marshall, T.A. Lachlan-Cope, T. Bracegirdl, T. Maksym, M.P. Meredith, Z. Wang and A. Orr, *Non-annular atmospheric circulation change induced by stratospheric ozone depletion and its role in the recent increase of Antarctic sea ice extent*, Geophysical Research Letters, 36 (2009)

[2-12]   S. Rahmstorf and H.J. Schellnhuber, *Der Klimawandel*, C.H.Beck, ISBN 978-3-406-63385-0 (2012)

[2-13]   R.M. DeConto, S. Galeotti, M. Pagani, D. Tracy, K. Schaefer, T.J. Zhang, D. Pollard and D.J. Beerling, *Past extreme warming events linked to massive carbon release from thawing permafrost*, Nature, 484 (7392) pp. 87-91 (2012)

[2-14]   M. Milankovich (ed.), *Mathematische Klimalehre und astronomische Theorie der Klimaschwankungen*, Bornträger, Berlin (1930)

[2-15]   R.A. Muller and G.J. MacDonald, *Spectrum of 100 kyr Glacial Cycle:orbital inclination, not eccentricity*, Proc. Natl. Acad.Sci. USA, 94, 8329-8334 (1997) http://muller.lbl.gov/papers/nas.pdf

[2-16]   K. Schaefer, T. Zhang, L. Bruhwiler and A.P. Barrett, *Amount and timing of permafrost carbon release in response to climate warming*, Tellus B, 63, 165-180 (2011)

[2-17]   N. Oreskes, *Beyond the ivory tower – The scientific consensus on climate change*, Science 306, 1686 (2004)

[2-18]   GRS (Gesellschaft für Anlagen- und Reaktorsicherheit), *Sicherheitsanalyse für Siedewasserreaktoren*, ISBN 3-92 38 75-52-5 (June 1993)

[2-19]   J. Knebel, C. Fazio, W. Maschek and W. Tromm, *Was tun mit dem nuklearen Abfall*, Spektrum der Wissenschaften, p. 34-41 (Feb 2013)

[3-1]   E. Mills, *The 230-billion Global Lighting Energy Bill*, International Association for Energy-Efficient Lighting and Lawrence Berkeley National Laboratory, http://evanmills.lbl.gov/pubs/pdf/global_lighting_energy.pdf, (June 2002)

[4-1]   D.M. Dodge, *Illustrated History of Wind Power Development*, (2001) http://telosnet.com/wind/

[4-2]   www.ewea.org

[4-3]   http://hydroquebec.com/learning/eolienne

[4-4]   GWEC (Global Wind Energy Council, 2010), *Global Wind Report 2010 and 2006*, http://gwec.net

[4-5]   EWEA (European Wind Energy Association, 2012), *Wind in Power – 2011 European Statistics*, http://www.ewea.org/fileadmin/files/library/publications/statistics/Wind_in_power_2011_European_statistics.pdf

[4-6]   BTM Wind Market Report (July 2010) http://www.renewableenergyworld.com

[4-7]   Roland Berger (2009), *Wind Energy Manufacturers challenges*, www.rolandberger.com

[4-8]   M. Ragheb, *Solar Thermal Power and Energy Storage Historical Perspective* (08/11/2011) https://netfiles.uiuc.edu/mragheb/NPRE%20498ES%20Energy%20Storage

[4-9]   IEA SHC (International Energy Agency Solar Heating and Cooling Program), Technical note: *Recommendation: Converting solar thermal collector area into installed capacity (m² to kW$_{th}$)*, available from www.iea-shc.org

[4-10]   IEA, *Converting Installed Solar Collector Area & Power Capacity into Estimated Annual Solar Collector Energy Output*, http://www.iea-shc.org/Data/Sites/1/document

[4-11]   Bank Sarasin, Solarstudie (2011)

[4-12]   WBGU (German Advisory Council on Global Change) (2009), *Future Bioenergy and Sustainable Land Use*, ISBN 978-84407-841-7, also in http://www.wbgu.de/fileadmin/templates/dateien/veroeffentlichungen/hauptgutachten/jg2008/wbgu_jg2008_en.pdf

[4-13]   A. Faaij, *Bioenergy and global food security*, ISBN 978-3-939 6191-21-9 (2008)

[4-14]   Prof. Adolf Götzberger, personal communication (2005)

[4-15]   BP, *Statistical review of world energy 2012*, http://www.bp.com/assets/bp_internet/globalbp/globalbp_uk_english/reports_and_publications/statistical_energy_review_2011/STAGING/local_assets/pdf/statistical_review_of_world_energy_full_report_2012.pdf

[4-16]   H-B. Horlacher, *Globale Potenziale der Wasserkraft*, externe Expertise für das WBGU-Hauptgutachten 2003 „Welt im Wandel: Energiewende zur Nachhaltigkeithttp://www.wbgu.de/fileadmin/templates/dateien/veroeffentlichungen/hauptgutachten/jg2003/wbgu_jg2003_ex03.pdf

[4-17]   IEA Energy Technology Systems Analysis Programme (ETSAP), *Hydropower*, (2010), http://www.iea-etsap.org/web/e-techds/pdf

[4-18]   The Green Age, *Rance Tidal Power Station*, France, http://www.thegreen-age.co.uk/greencommercial/tidal-energy/rance-tidal-power

[4-19]   BMU, *Erneuerbare Energien: Innovationen für die Zukunft* (2004), http://www.dlr.de/Portaldata/41/Resources/dokumente/institut/system/publications/broschuere_ee_innov_zukunft.pdf

[4-20]   DLR, *Innovationen für die Zukunft*, (Mai 2004) http://www.dlr.de/Portaldata/41/Resources/dokumente/institut/system/publications/broschuere_ee_innov_zukunft.pdf

[5-1]   O. Ristau, *The Photovoltaic Market in Japan: unquestioned Leadership of World Market*, http://www.solarserver.com/solarmagazin/artikelseptember2001, (15.9.2001)

[5-2]   Press Release BSW (12.01.2012) Solar, *Weiterer Solarstrom-Ausbau erhöht Strompreis kaum noch*, http://www.solarwirtschaft.de/presse-mediathek/pressemeldungen-im-Detail/

[5-3]   M. Braun, T. Degner, T. Glotzbach and Y.-M. Saint-Drenan, *Wertigkeit von PV-Strom - Nutzen durch Substitution des konventionellen Kraftwerkparks und verbrauchsnahe Erzeugung*, OTTI Tagungsband, 23rd Symposium Photovoltaische Solarenergie, pp. 43-48, ISBN: 978-3-934681-67-5 (2008)

[5-4]   LBBW (02/2008) *Valuing the invaluable*, Market model 3.0

[5-5]   SüdWestStrom, *Marktstudie zur Strompreisentwicklung in Deutschland, Vorgehensweise und indikative Ergebnisse einer Berechnung*, (Mai 2008)

[6-1]   D.M. Chapin, C.S. Fuller and G.L. Pearson, *A New Silicon p-n Junction Photocell for Converting Solar Radiation into Electrical Power*, Bell Telephone Laboratories, Inc., Murray Hill (Jan 1954)

[6-2]   B. Prior and C. Campbell, GTM Research, *Polysilicon 2012-2016: Supply, Demand & Implications for the Global PV Industry*, (2012) http://www.greentechmedia.com/research/report/polysilicon-2012-2016

[6-3]   BP patent (20.01.2006), US 8048221 B1

[6-4]   Matthias Müller, SCHOTT AG, personal communication (2013)

[6-5]   Wim Sinke, ECN Solar Energy, personal communication (2013)

[6-6]   A. Metz, D. Adler, S. Bagus, H. Blanke, M. Bothar, E. Brouwer, S. Dauwe, K. Dressler, R. Droessler, T. Droste, M. Fiedler, Y. Gassenbauer, T. Grahl, N. Hermert, W. Kuzminskib, A. Lachowicz, T. Lauinger, N. Lenck, M. Manole, M. Martini, R. Messmer, C. Meyer, J. Moschner, K. Ramspeck, P. Roth, R. Schoenfelder, B. Schum, J. Sticksel, K. Vaas, M. Volk and K. Wangemann, *Industrial high performance silicon solar cells and modules based on rear surface passivation technology*, Solar Energy Materials and Solar Cells, 120, Part A, pp. 417-425 (2014)

[6-7]   Joachim Luther, SERIS data, personal communication (2013)

[7-1]   H. Ebbinghaus, *Über das Gedächtnis. Untersuchungen zur experimentellen Psychologie*. Leipzig: Duncker & Humblot, (1885)

[7-2]   T.P. Wright, *Factors affecting the cost of air planes*, Journal of Aeronautical Sciences 3(4), 122-128 (1936)

[7-3]   B. Henderson, *The experience Curve – Reviewed*, The Boston Consulting Group, Perspective N° 124 (1974)

[7-4]   Internal information on DRAM, Applied Materials, Semiconductor (2009)

[7-5]   Internal information on Flat Panel Display, Applied Materials, Display (2009)

[7-6]   Internal information on coated and architectural glass, Applied Materials, Energy and Environment Solutions (2009)

[7-7]   SEMI, *International Technology Roadmap for Photovoltaics (ITRPV)*, www.itrpv.net (2013)

[7-8]   W. Hoffmann, S. Wieder and T. Pellkofer, *Differentiated Price Experience Curves as Evaluation Tool for Judging the Further Development of Crystalline and Thin Film PV Solar Electricity Products*, 24th PVSEC Hamburg (2009)

[8-1]   R. Brendel, personal communication (2013)

[8-2]   E. Lantz, M. Hand and R. Wiser, *The Past and Future Cost of Wind Energy*, http://www.nrel.gov/docs/fy12osti/54526.pdf (August 2012)

[8-3]   S. Schoenung, *Energy Storage Systems Cost Update, A Study for the DOE Energy Storage Systems Program*, SANDIA report, SAND2011-2730 (online ordering http://www.osti.gov/bridge) (2011)

[8-4]   J. Holman, *Increasing Transmission Capacity*, http://windsystemsmag.com/article/detail/191/increasing-tranmission-capacity (January 2011)

[8-5]   V. Quaschning, *Systemtechnik einer klimaverträglichen Elektrizitätsversorgung in Deutschland für das 21. Jahrhundert*, VDI-Verlag Düsseldorf 2000, ISBN: 3-318.343706-6 (2000), http://www.volker-quaschning.de/klima2000/Kapitel4.html

[9-1]   IEA, World Energy Outlook (WEO) (2002)

[9-2]   E.U. von Weizsäcker, A.B. Lovins and L.H. Lovins, *Factor Four: Doubling Wealth – Halving Resource Use: The New Report to the Club of Rome*, Earthscan, London (1998)

[9-3]   E.U. von Weizsäcker, K. Hargroves and M. Smith (2009), *Factor Five. Transforming the Global Economy through 80% Improvements in Resource Productivity*, Earthscan, London

[9-4]   German Advisory Council on Global Change (WBGU), *World in Transition – towards sustainable energy systems*, Earthscan, ISBN 3-540-40160-1 (2003)

[9-5]   Greenpeace, *Energy [R]evolution 2012*, www.greenpeace.org

[9-6]   J. Randers, *2052. A Global Forecast for the Next Forty Years*, Chelsea Green Publishing, White River Junction/Vermont, USA, ISBN 978-3-86581-398-5 (2012)

[9-7]   United Nations, Department of Economic and Social Affairs, Population Division, *World Population Prospects: The 2012 Revision, Highlights and Advance Tables*. Working Paper No. ESA/WP.228 (2013)

[9-8]   J. Nitsch and J. Luther, *Energieversorgung der Zukunft*, Springer Verlag, ISBN 3-540-51753-7 (1990)

[9-9]   H.M. Henning and A. Palzer, *A comprehensive model for the German electricity and heat sector in a future system with a dominant contribution from renewable energy technologies*, renewable & sustainable energy reviews, ISSN: 1364-0321 (2013)

[10-1]   Nicholas Stern (2006), *Stern Review on the Economics of Climate Change"*, http://www.hm-treasury.gov.uk

[10-2]   J. Verne (1875), *Die geheimnisvolle Insel*, ISBN 978-3-401-00260-6

[10-3]   J. O.M. Bockris, E. W. Justi, *Wasserstoff. Energie für alle Zeiten. Konzept einer Sonnen-Wasserstoff-Wirtschaft.* Augustus Verlag, ISBN 3-8043-2591-2

[10-4]   L.A. Nefiodow, *Wege zur Produktivität und Vollbeschäftigung im Zeitalter der Information*, 3. Auflage, ISBN 398 051 4439 (1999)

[10-5]   W.R. Cline, *Comments on the Stern Review*, http://www.iie.com/publications/papers/paper/cfm?ResearchID=874 (2008)

Index